Radiotherapy and Brachytherap

NATO Science for Peace and Security Series

This Series presents the results of scientific meetings supported under the NATO Programme: Science for Peace and Security (SPS).

The NATO SPS Programme supports meetings in the following Key Priority areas: (1) Defence Against Terrorism; (2) Countering other Threats to Security and (3) NATO, Partner and Mediterranean Dialogue Country Priorities. The types of meeting supported are generally "Advanced Study Institutes" and "Advanced Research Workshops". The NATO SPS Series collects together the results of these meetings. The meetings are co-organized by scientists from NATO countries and scientists from NATO's "Partner" or "Mediterranean Dialogue" countries. The observations and recommendations made at the meetings, as well as the contents of the volumes in the Series, reflect those of parti-cipants and contributors only; they should not necessarily be regarded as reflecting NATO views or policy.

Advanced Study Institutes (ASI) are high-level tutorial courses intended to convey the latest developments in a subject to an advanced-level audience

Advanced Research Workshops (ARW) are expert meetings where an intense but informal exchange of views at the frontiers of a subject aims at identifying directions for future action

Following a transformation of the programme in 2006 the Series has been re-named and re-organised. Recent volumes on topics not related to security, which result from meetings supported under the programme earlier, may be found in the NATO Science Series.

The Series is published by IOS Press, Amsterdam, and Springer, Dordrecht, in conjunction with the NATO Public Diplomacy Division.

Sub-Series

A.	Chemistry and Biology	Springer
B.	Physics and Biophysics	Springer
C.	Environmental Security	Springer
D.	Information and Communication Security	IOS Press
E.	Human and Societal Dynamics	IOS Press

http://www.nato.int/science
http://www.springer.com
http://www.iospress.nl

Series B: Physics and Biophysics

Radiotherapy and Brachytherapy

edited by

Yves Lemoigne
European Scientific Institute
Site d'Archamps
France

and

Alessandra Caner
European Scientific Institute
Site d'Archamps
France

 Springer

Published in cooperation with NATO Public Diplomacy Division

Proceedings of the NATO Advanced Study Institute on
Physics of Modern Radiotherapy & Brachytherapy Archamps
France
15–27 November 2007

Library of Congress Control Number: 2009931108

ISBN 978-90-481-3096-2 (PB)
ISBN 978-90-481-3095-5 (HB)
ISBN 978-90-481-3097-9 (e-book)

Published by Springer,
P.O. Box 17, 3300 AA Dordrecht, The Netherlands.

www.springer.com

Printed on acid-free paper

CONTENTS

PREFACE

This book reports the majority of lectures given during the NATO Advanced Study Institute ASI-982996, which was held at the European Scientific Institute of Archamps (ESI, Archamps – France) from November 15 to November 27, 2007. The ASI course was structured in two parts: the first was dedicated to what is often called "teletherapy", i.e. radiotherapy with external beams, while the second focused on internal radiotherapy, also called "brachytherapy" or "curietherapy" in honour of Madame Curie who initiated the technique about a century ago. This ASI took place after the European School of Medical Physics, which devoted a 3 week period to medical imaging, a subject complementary to the topics of this book. Courses devoted to nuclear medicine and digital imaging techniques are collected in two volumes of the NATO Science Series entitled "Physics for Medical Imaging Applications" (ISBN 978-1-4020-5650-5) and "Molecular imaging: computer reconstruction and practice" (ISBN 978-1-4020-8751-6).

Every year in autumn ESI organises the European School of Medical Physics, which covers a large spectrum of topics ranging from Medical Imaging to Radiotherapy, over a period of 5 weeks. Thanks to the Cooperative Science and Technology sub-programme of the NATO Science Division, weeks four and five were replaced this year by the ASI course dedicated to "Physics of Modern Radiotherapy & Brachytherapy". This allowed the participation of experts and students from 20 different countries, with diverse cultural background and pro-fessional experience. A further positive outcome of NATO ASI participation is the publication of this book, containing the lecture series contributed by speakers during the ASI. We hope it will be a reference book in radiotherapy, addressing an audience of young medical physicists everywhere in the world, who are wishing to review the physics foundations of the relevant technologies, catch up with the state of the art and look ahead into future developments in their field, e.g. the emerging technique of "Hadrontherapy".

This volume starts with an introductory lecture about radiotherapy strategy and accuracy requirement, two aspects which are of the first importance (G. Hartmann). Fundamental radiobiology is reviewed by Francesca Buffa, fol-lowed by G. Hartmann discussing photon and electron dosimetry. G. Hartmann also illustrates the nomenclature in radiotherapy treatment planning, specified in the ICRU-50 document. Three lectures by Steve Webb describe Conformal Radiotherapy (CRT), covering various modern techniques from Multi-Leaf Collimators (MLC) to Intensity Modulated RadioTherapy (IMRT).

Simeon Nill from DKFZ develops two topics: inverse planning and optimisation techniques first, followed by stereotactic treatment techniques. As well from DKFZ, Uwe Oelfke reviews image guided techniques in radiotherapy. Quality assurance in radiotherapy is everybody's deep concern and is discussed by A. McKenzie from Bristol.

Despite the quasi totality of patients being presently treated with gamma and X-rays, some novel powerful tools are emerging implementing proton and light ions (e.g. carbon ions) beams, which will progressively gain ground in patient treatment. It was deemed important to introduce this new approach, going under the comprehensive name of Hadrontherapy (protons and light ions, e.g. Carbon ions, all belong to a group of particles called hadrons.). Ugo Amaldi presents an overview of the field, whereas U. Oelfke focuses on proton therapy and its implementation at the Heidelberg Hadron Therapy centre. The role of medical imaging in radiotherapy (clinical applications and geometrical uncertainties) is presented by Peter Remeijer from Amsterdam. Finally, the chief of the radiotherapy service of the "HCUGE Hospital" concludes this section with two lectures devoted to clinical application of 3D conformal radiotherapy.

In the Brachytherapy section, a review of the sources and of the conventional computer calculations used in this field is presented by A. Rijnders (Brussels). Among different dosimetry systems for interstitial brachytherapy, Ginette Marinello privileges a description of the so-called "Paris System". Cervix cancer treatment is then analysed in details by Inger-Lena Lamm (Lund Sweden). Patient dose calculation using modern Monte-Carlo techniques is presented by J. Perez-Calatayud from Valencia (Spain).

We wish to thank all the participants, who allowed the ASI at Archamps to be a success within an excellent international atmosphere: lecturers, students (who participated actively) and all the ESI team (Manfred Buhler-Broglin, Alessandra Caner, Violaine Presset, Filiz Demolis and Julien Lollierou).

Many thanks to the Hôpital Cantonal de Genève, which hosted us twice (Philippe Nouet) and gave us access to radiotherapy and brachytherapy services (Dr. Miralbell and Dr. Popowski).

Finally, we wish to thank and express the gratitude to the Cooperative Science and Technology sub-programme of the NATO Science Division, lead by Prof. Fausto Pedrazzini, without whom this Advance Study Institute would have not been possible.

Yves Lemoigne[1]
Co-Director of ASI-982440

---

[1] European School of Medical Physics, European Scientific Institute, Bâtiment Le Salève, Site d'Archamps F-74166 Archamps (France).

LIST OF PARTICIPANTS

Lecturers' surnames are in bold

Surname	Name	Country	City	E-mail address
Abate	Armando	Italy	Rome	armando.abate@yahoo.it
Acar	Hilal	Turkish	Istanbul	hilalacar@hotmail.com
Amaldi	Ugo	Italy	Milano	ugo.amaldi@cern.ch
Avitabile	Rossella	Italy	Padova	rossella.avitabile@libero.it
Balabanova	Anna	Bulgaria	Sofia	abalabanova@gmail.com
Bartesagi	Giacomo	Italy	Milan	giacomo.bartesaghi@unimi.it
Bonnet	Jacques	France	Toulouse	bonnet.jacques@claudiusregaud.fr
Boubir	Messaoud	Algeria	Batna	boubir_mess@yahoo.fr
Bourbia	Nadjla	Algeria	Jijel	nj-br@hotmail.com
Bucciarelli	Bernard	France	Sevran	bbi@nucletron.fr
Buffa	Francesca	UK	Oxford	francesca.buffa@imm.ox.ac.uk
Caner	Alessandra	Italian	Archamps	alecaner@gmail.com
Carrasco	Fatima	Portugal	Porto	fcarrasco@fis.ua.pt
Ceric	Melisa	BiH	Sarajevo	mceric@epn.ba
Conti	Valentina	Italy	Milano	conti.valentina@gmail.com
Djarova	Anna	Bulgaria	Sofia	anna.djarova@gmail.com
Dogandjiiska	Daniela	Bulgaria	Sofia	d.doganjiiska@abv.bg
Demolis	Filiz	Turkey	Izmir	filiz@esi.cur-archamps.fr
Derkaoui	Jamal	Morocco	Oujda	derkaoui@sciences.univ-oujda.ac.ma
Dries	Wim	Netherlands	Eindhoven	wim.dries@catharina-ziekenhuis.nl
Drolc	Botajan	Slovenia	Ljubljana	bostajn.drolc@zvd.si
Dzharova	Anna	Bulgaria	Sofia	anna.djarova@gmail.com
Dumitriu	Irina	Romania	Bucharest	fiorrdaliss@gmail.com
Falie	Dragos	Romania	Bucharest	dfalie@alpha.imag.pub.ro
Ferrand	Regis	France	Orsay	regis.ferrand@curie.net
Fezov	Daniel	Bulgaria	Sofia	daniel_fezov@abv.bg
Fekete	Zsolt	Romania	Cluj	drfekete@gmail.com

(continued)

LIST OF PARTICIPANTS (continued)

Georgiev	Ivaylo	Bulgaria	Sofia	iv3georg@yahoo.com
Ghitulescu	Zoe	Romania	Bucharest	ghitulescuzoe@yahoo.com
Goksel	Evren Ozan	Turkey	Istanbul	evrenozangoksel@yahoo.com
Hamdy Abdelmaboud	Ahmed	Egypt	Cairo	Benhamdy@yahoo.com
Hartmann	Gunther	Germany	Heidelberg	g.hartmann@dkfz-heidelberg.de
Hristova	Julia	Bulgaria	Sofia	jully_84@mail.bg
Ivanova	Severina	Bulgaria	Sofia	sevi_ii@yahoo.com
Jarry	Patrick	France	Gif-sur-Yvette	patrick.jarry@cern.ch
Jobs	J. T.	Germany	Freiburg	ruediger.lauk@ptw.de
Kaufhold	Alexander	Germany	Freiburg	alexander.kaufhold@ptw.de
Kaulich	Theodor	Germany	Tübingen	theodor.kaulich@med.uni-tuebingen.de
Kostova-Lefterova	Desislava	Bulgaria	Sofia	dessi.zvkl@gmail.com
Lafrenière	Geneviève	UK	Yarnton	Genevieve.Lafreniere@varian.com
Lamm	Inger-Lena	Sweden	Lund	Inger-lena.lamm@skane.se
Lanche	Stephanie	France	Archamps	lanchestephanie@gmail.com
Lemoigne	Yves	France	Archamps	psf1@orange.fr
Li	Pan	Germany	Heidelberg	pan.li@dkfz.de
Loja Nunes Pires	Paola	Germany	Berlin	paola.loja@physik.fu-berlin.de
Lollierou	Julien	France	Presilly	jlollierou@yahoo.fr
Marinello	Ginette	France	Creteil	ginette.marinello@hmn.ap-hop-paris.fr
Mazeron	J-Jacques	France	Paris	jean-jacques.mazeron@psl.ap-hop-paris.fr
McKenzie	Alan	UK	Bristol	alan.mckenzie@ubht.swest.nhs.uk
Marinov	Filip	Bulgaria	Sofia	filip_marinov@abv.bg
Masiuk	Mariusz	Poland	Gdańsk	masiuk@amg.gda.pl
Miralbell	Raymond	Spain	Barcelona	raymond.miralbell@hcuge.ch
Mukov	Mihail	Bulgaria	Sofia	mmukov@gmail.com
Multan	Marta	Poland	Krakow	multan@if.uj.edu.pl
Nikollari	Ermal	Albania	Tirana	ermal.nikollari@libero.it

(continued)

Nikolov	Ivaylo	Bulgaria	Sofia	ivo@odigy.com
Nill	Simeon	Germany	Heidelberg	s.nill@dkfz-heidelberg.de
Nouet	Philippe	Switzerland	Genève	philippe.nouet@hcuge.ch
Oelfke	Uwe	Germany	Heidelberg	u.oelfke@dkfz-heidelberg.de
Pasciuti	Katia	Italy	Rome	ka.pasciuti@libero.it
Perez-Calatayud	José	Spain	Valencia	jose.perez-calatayud@uv.es
Popowski	Youri	Belgium	Brussels	Youri.popowski@hcuge.ch
Remeijer	Peter	Netherlands	Amsterdam	Prem@nki.nl
Rijnders	Alex	Belgium	Brussels	a.rijnders@europehospital.be
Ruiz-Bueno	Antonio	Spain	Valladolid	aruizbueno@gmail.com
Samet	Nina	Moldova	Chisinau	nina_smd@yahoo.com
Saoud	Houda	Morocco	Rabat	saoudhanaa@yahoo.fr
Shekel	Efrat	Israel	Petach Tikva	efrat.shekel@gmail.com
Smail	Hassen	Algeria	Batna	smail_hassen@yahoo.fr
Toschi	Nicola	Italy	Rome	toschi@med.uniroma2.it
Webb	Steve	UK	London	steve@icr.ac.uk
Zajacova	Zuzana	Slovakia	Bratislava	zuzana.zajacova@cern.ch

PHOTOGRAPHS

Class room

Computer room

G. Hartmann (De) lecturing

PC exercises with G. Hartmann

Working group

Equipement demonstration

Equipment explanation

Steve Webb (UK) Lecturing

Work at Geneva hospital

J. Bonnet (Fr) lecturing

Wim Dries (Nl) lecturing

T. Kaulich (De) lecturing

G. Marinello (Fr) lecturing

Tutorial on Brachytherapy

I. Lamm (Se) at Geneva Hospital

After lunch Coffee

Farewell dances

Farewell Dinner

Advanced Study Institute CBP.ASI 982996 - Archamps (Fr)

PART I:

INTRODUCTION TO RADIOTHERAPY

FUNDAMENTAL RADIOBIOLOGY AND ITS APPLICATION TO RADIATION ONCOLOGY

FRANCESCA M. BUFFA
Weatherall Institute of Molecular Medicine,
University of Oxford, Oxford, UK
Francesca buffa@imm.ox.ac.uk.

Abstract A brief overview of fundamental concepts in radiobiology is provided. The concept of cell survival as ability to retain reproductive integrity is introduced, and critical variables influencing the cell survival curve as a function of absorbed radiation dose are discussed. Application of these concepts to radiation oncology and radiotherapy is then outlined. Examples are provided of clinical studies that can be performed with current high-throughput molecular biology and imaging technology. The type of information derived from these studies and its potential are discussed in the context of radiotherapy trial design, radiotherapy schedule and modality tailoring, and planning of treatment dose.

Keywords: DNA damage and repair; cell survival curves; tumour control; morbidity; molecular oncology; radiation oncology; radiotherapy

1. Radiation effects and radiobiological modeling

1.1. DNA DAMAGE AND REPAIR

A variety of lesions are produced in the cell following irradiation. However, the damage most critical for determining cell survival, and thus the most relevant to standard external radiotherapy, is that created in the target molecule DNA (for a review see G.G. Steel, 2002[1]). The lesions in DNA produced by irradiation include base damage, single strand breaks and double strand breaks. Once double strand breaks are produced, several scenarios may occur, with damage either being repaired accurately or inaccurately. Specifically, breaks may rejoin in their original configuration (repair), may fail to rejoin (deletion), or broken ends

Y. Lemoigne and A. Caner (eds.), *Radiotherapy and Brachytherapy*,
© Springer Science + Business Media B.V. 2009

may rejoin other broken ends (mis-repair). Mis-repaired double strand breaks can yield various types of chromosome or chromatid aberrations, and generally have the greatest detrimental effect on cell survival and division.

Radiation can induce DNA ionization by direct or indirect mechanisms.[1] Direct action involves direct interaction of radiation with DNA, and high-linear energy transfer (high-LET) radiation produces most damage by this route. In contrast, indirect action occurs when radiation interacts initially with another molecule, usually water, producing free radicals which then diffuse to the DNA. These free radicals can then react with oxygen species thus fixing the DNA damage. In the presence of oxygen, most of the damage from low-linear energy transfer (low-LET) radiation is due to this indirect type of action. As many tumors are hypoxic, compounds that mimic oxygen such as nitroimidazoles can improve tumor control during low-LET radiotherapy.

1.2. CELL SURVIVAL CURVES AND RADIOBIOLOGICAL MODELLING

In radiobiology, cell survival is defined as the ability to form a colony of progeny. In other words, a surviving cell is a cell that retains its *reproductive integrity*, and this can be measured by testing for *clonogenicity*. However, even lethally irradiated cells can undergo residual cell divisions. Therefore, it is customary in clonogenic assays to count colonies only if they contain at least 50 cells, which corresponds to at least 5 cell divisions. In these assays the ratio between colony-forming efficiency of one irradiated cell population and that of a non-irradiated control population is determined. Resulting cell survival curves show the relationship between absorbed dose and the fraction of cells retaining their reproductive integrity. Conventionally, surviving fraction is plotted on a logarithmic scale on the ordinate against dose on the abscissa. As dose increases the surviving fraction of cells (SF) decreases exponentially. In the case of low-LET, SF is approximately equal to the inverse of a linear-quadratic exponential function of dose. As LET increases the linear exponential component tends to dominate the curve. Furthermore, the dose-rate affects the shape of the SF curve; specifically, at low dose-rates the SF after the same total absorbed dose is higher and the quadratic component of the exponential curve is reduced. Several models have been proposed to describe the SF curve and the mechanisms controlling it (for reviews see e.g. E.L. Alpen[2]; R.K. Sachs et al.[3]), of which the most widely used to date is the linear-quadratic (LQ) model. However, at low doses SF has been shown to depart significantly from this model (for a review see M.C. Joiner et al.[4]).

1.2.1. *The linear-quadratic (LQ) model*

In the linear-quadratic (LQ) model the SF is described as a function of dose using the formula:

$$SF = e^{-\alpha \cdot D - \beta \cdot D^2}$$

where D is the dose, and α and β are radiosensitivity parameters which can be estimated by fitting cell survival data. The initial shape of SF curves is determined by α, whereas β is responsible for the shape at higher doses. The LQ model is a semi-empirical model, and as discussed above several mechanistic approaches have been developed. However, its simplicity of expression and relative robustness at intermediate dose ranges has made it attractive for clinical applications. The α and β parameters have been given various interpretations[3] and the initial model expression has been extended to account for factors such as proliferation during prolonged irradiation, DNA damage repair[5-7] (for a review see G.G. Steel[1]), hypoxia and reoxygenation (e.g. D.J. Brenner et al.[8]) and low-dose hypersensitivity.[4]

1.2.2. *Tumour control and normal tissue complication probability, and the 4-Rs of radiotherapy*

Several parameters influence the radiotherapy dose-response curve in tumors and normal tissues. The LQ model has been used extensively as a component of more complicated models to estimate the probabilities of tumor control after radiotherapy (tumor control probability, *tcp*), and of damage to normal surrounding tissue, normal tissue complication probability (*ntcp*). The parameters that have been incorporated to date into these models include radiosensitivity (or the α/β ratio), its variation among the population,[9-11] volume effects,[12, 13] and the so-called "4-Rs of radiotherapy"[14]:- tumor repopulation, DNA damage repair, cell cycle redistribution and re-oxigenation.[15, 16] The relative importance of these parameters and their influence on tumour control will depend on the length of the radiotherapy schedule, the dose-rate of the radiotherapy modality, the hypoxic status of the tumour and the tumour type. The time-of-appearance and the dose-response curve of complications in normal tissue depend also upon the proliferative organization of that tissue.[17-19] Although clinical data are few, and the range of doses and fractionation schemes are limited, an increasing number of studies demonstrate that the dose-response relationship can be estimated from retrospective information on radiotherapy treatment success and failure. This has prompted attempts to estimate biological parameters characterizing these dose-response curves.[20–22] and to correlate surviving fraction in colony-forming assays with differences in radiosensitivity between tumours.[23–26]

2. Radiobiology and radiation oncology

2.1. PLANNING OF RADIOTHERAPY TRIALS

Radiotherapy dose-response curves are useful for estimating the sample size required to resolve any potential effect of a given dose modification, and are therefore important when planning radiotherapy trials.[27] Intuitively, detection of any dose-dependent effect will be easiest when the dose-response curve is steepest, and thus the number of patients needed to see an effect will be lower. Whilst trials are often correctly powered to observe the effect of a given dose modification on tcp, they tend to be under-powered for detection of morbidity changes. This is due to a combination of factors. Firstly, morbidity in radiotherapy trials is usually low and this means that the curve of morbidity versus dose is not very steep. Thus a large sample size is needed to resolve the effect on morbidity of a given dose modification. Furthermore, morbidity, and particularly that occurring at a late time point, tends to be under-reported in clinical trials. However, there is an increasing effort towards incorporating morbidity into optimisation of radiotherapy. Both separate estimation of trial size for local control and morbidity and the use of a single combined endpoint incorporating both local control and morbidity information.[27]

2.2. MOLECULAR ONCOLOGY AND RADIATION ONCOLOGY

Recent technological progress, coupled with advances in molecular and cancer biology, has dramatically improved our ability to quantify a large number of biological and physiological characteristics of normal human tissue and tumors. An increasing number of molecular oncology studies measure variables such as protein and gene expression *ex-vivo*. In radiation oncology, such studies enable translation of knowledge from laboratory to clinical studies and vice-versa, so strengthening understanding of *in-vivo* radiobiology of tumors and their interaction with specific treatment schedules.[28–34] Specifically, these studies could help to improve understanding of radiotherapy treatment failure and complications, enable tailored patient-specific radiotherapy treatment, and assist discovery of potential novel targets for combined modality therapies. For example, direct activation of the cell surface epidermal growth factor receptor (EGFR) by radiation leads to increased proliferation and enhanced radioresistance and EGFR expression has been associated with accelerated repopulation following irradiation.

Thus EGFR presents both an interesting molecular marker for tailoring radio-therapy schedule and target for complementary biological therapy. Clinical and laboratory studies have provided evidence that expression of EGFR in *ex-vivo* samples is linked to tumour repopulation after radiotherapy, and that high EGFR expression can potentially identify patients who would benefit from accelerated radiotherapy.[35, 36]

2.3. RADIOBIOLOGY AND TREATMENT PLANNING

Radiobiological and biological knowledge gained from parallel clinical and laboratory studies, together with therapeutic ratio information, might also be used to optimise different radiotherapy dose and dose-distribution plans. Some pioneering works are validating the feasibility of using new imaging and molecular biology techniques to inform therapy decisions by providing maps of tumor biology which can be translated into radiation dose maps.[37] However, the relative paucity of accurate information regarding the radiobiology, biology and microenvironmental dynamics within whole tumors has so far greatly limited the accuracy and therefore clinical application of these studies.[38] Further complications arise when considering that in the clinical context radiotherapy is often combined with chemotherapy. Some research studies are now beginning to incorporate either drug-alone effects or combined drug-radiation effects into radiation dose-response models,[39-41] but these models will need to be validated in large clinical studies.

References

1. G.G. Steel: Basic Clinical Radiobiology, Oxford University Press, 3rd Edition, 2002.
2. E.L. Alpen: Radiation Biophysics, Prentice-Hall, International Editions, London, 1990.
3. R.K. Sachs, P. Hahnfeld, D.J. Brenner: The link between low-LET dose-response relations and the underlying kinetics of damage production/repair/misrepair, Int J Radiat Biol., 72(4):351–374, 1997.
4. M.C. Joiner, B. Marples, P. Lambin, S.C. Short, I. Turesson: Low-dose hypersensitivity: current status and possible mechanisms, Int J Radiat Oncol Biol Phys., 49(2):379–389, 2001.
5. R.G. Dale: The application of the linear quadratic dose effect equation to fractionated and protracted radiotherapy, Br J Radiol., 58:515–528, 1985.
6. H.D. Thames, J.H. Hendry: Fractionation in Radiotherapy, Taylor & Francis, London, 1987.
7. J.F. Fowler: The linear-quadratic formula and progress in radiotherapy, Br J Radiol., 62:679–694, 1989.
8. D.J. Brenner, L.R. Hlatky, P.J. Hahnfeldt, E.J. Hall, R.K. Sachs: A convenient extension of the linear-quadratic model to include redistribution and reoxygenation, Int J Radiat Oncol Biol Phys., 32(2):379–390, 1995.

9. S.M. Bentzen: Steepness of the clinical dose-control curve and variation in the in vitro radiosensitivity of head and neck squamous cell carcinoma, Int J Radiat Biol., 61(3):417–423, 1992.

10. H. Suit, S. Skates, A. Taghian, P. Okunieff, J.T. Efird: Clinical implications of heterogeneity of tumor response to radiation therapy, Radiother Oncol., 25(4):251–260, 1992.

11. S. Webb, A.E. Nahum: A model for calculating tumour control probability in radiotherapy including the effects of inhomogeneous distributions of dose and clonogenic cell density, Phys Med Biol., 38(6):653–666, 1993.

12. D.J. Brenner: Dose, volume, and tumor-control predictions in radiotherapy, Int J Radiat Oncol Biol Phys., 26(1):171–179, 1993.

13. S.M. Bentzen, H.D. Thames: Tumor volume and local control probability: clinical data and radiobiological interpretation, Int J Radiat Oncol Biol Phys., 1; 36(1):247–251, 1996.

14. H.R. Withers: The four R's of radiotherapy, Adv Radiat Biol., 5:241–247, 1975.

15. Tucker SL, Taylor JM: Improved models of tumour cure. Int J Radiat Biol., 70(5):539–553, 1996.

16. M. Zaider, G.N. Minerbo: Tumour control probability: a formulation applicable to any temporal protocol of dose delivery, Phys Med Biol., 45(2):279–293, 2000.

17. A. Niemierko, M. Goitein: Calculation of normal tissue complication probability and dose-volume histogram reduction schemes for tissues with a critical element architecture, Radiother Oncol., 20(3):166–176, 1991.

18. C. Burman, G.J. Kutcher, B. Emami, M. Goitein: Fitting of normal tissue tolerance data to an analytic function, Int J Radiat Oncol Biol Phys., 21(1):123–135, 1991.

19. A. Jackson, G.J. Kutcher, E.D. Yorke: Probability of radiation-induced complications for normal tissues with parallel architecture subject to non-uniform irradiation. Med Phys., 20(3):613–625, 1993.

20. S.M. Bentzen, L.V. Johansen, J. Overgaard, H.D. Thames: Clinical radiobiology of squamous-cell carcinoma of the oropharynx, Int J Radiat Oncol Biol Phys., 20(6):1197–1206, 1991.

21. S. Webb: Optimum parameters in a model for tumour control probability including inter-patient heterogeneity, Phys Med Biol., 39(11):1895–1914, 1994.

22. J.D. Fenwick: Predicting the radiation control probability of heterogeneous tumour ensembles: data analysis and parameter estimation using a closed-form expression. Phys Med Biol., 43(8):2159–2178, 1998.

23. C.M. West, S.E. Davidson, S.A. Roberts, R.D. Hunter: The independence of intrinsic radio-sensitivity as a prognostic factor for patient response to radiotherapy of carcinoma of the cervix, Br J Cancer, 76(9):1184–1190, 1997.

24. T. Bjork-Eriksson, C. West, E. Karlsson, C. Mercke: Tumor radiosensitivity (SF2) is a pro-gnostic factor for local control in head and neck cancer, Int J Radiat Oncol Biol Phys., 46(1):13–19, 2000.

25. F.M. Buffa, S.E. Davidson, R.D. Hunter, A.E. Nahum, C.M.L. West: Incorporating biologic measurements (SF2, CFE) into a tumor control probability model increases their prognostic significance: A study in cervical carcinoma treated with radiation therapy, Int J Radiat Oncol Biol Phys., 50(5):1113–1122, 2001.

26. M. Baumann, M. Krause, R. Hill: Exploring the role of cancer stem cells in radioresistance, Nat Rev Cancer, 8(7):545–554, 2008.

27. S.M. Bentzen: Radiobiological considerations in the design of clinical trials, Radiother Oncol., 32(1):1–11, 1994.

28. J.H. Kaanders, K.I. Wijffels, H.A. Marres, A.S. Ljungkvist, L.A. Pop, F.J. van den Hoogen, P.C. de Wilde, J. Bussink, J.A. Raleigh, A.J. van der Kogel: Pimonidazole binding and tumor vascularity predict for treatment outcome in head and neck cancer, Cancer Res., 62:7066–7074, 2002.

29. F.M. Buffa, S.M. Bentzen, F.M. Daley, S. Dische, M.I. Saunders, P.I. Richman, G.D. Wilson: Molecular marker profiles predict locoregional control of head and neck squamous cell carcinoma in a randomized trial of continuous hyperfractionated accelerated radiotherapy, Clin Cancer Res., 10(11):3745–3754, 2004.

30. J.G. Eriksen, F.M. Buffa, J. Alsner, T. Steiniche, S.M. Bentzen, J. Overgaard: Molecular profiles as predictive marker for the effect of overall treatment time of radiotherapy in supraglottic larynx squamous cell carcinomas, Radiother Oncol., 72(3):275–282, 2004.

31. J. Overgaard, J.G. Eriksen, M. Nordsmark, J. Alsner, M.R. Horsman; Danish Head and Neck Cancer Study Group: Plasma osteopontin, hypoxia, and response to the hypoxia sensitiser nimorazole in radiotherapy of head and neck cancer: results from the DAHANCA 5 randomised double-blind placebo-controlled trial, Lancet Oncol., 6(10):757–764, 2005.

32. J. Akervall: Gene profiling in squamous cell carcinoma of the head and neck, Cancer Metastasis Rev., 24:87–94, 2005.

33. C.N. Andreassen: Can risk of radiotherapy-induced normal tissue complications be predicted from genetic profiles? Acta Oncol., 44:801–815, 2005.

34. M. Lobrich and J. Kiefer: Assessing the likelihood of severe side effects in radiotherapy, Int. J. Cancer, 118:2652–2656, 2006.

35. J.G Eriksen, T. Steiniche, J. Overgaard: The influence of epidermal growth factor receptor and tumor differentiation on the response to accelerated radiotherapy of squamous cell carcinomas of the head and neck in the randomized DAHANCA 6 and 7 study, Radiother. Oncol., 74:93–100, 2005.

36. S.M. Bentzen, B.M. Atasoy, F.M. Daley, S. Dische, P.I. Richman, M.I. Saunders, K.R. Trott, G.D. Wilson: Epidermal growth factor receptor expression in pretreatment biopsies from head and neck squamous cell carcinoma as a predictive factor for a benefit from accelerated radiation therapy in a randomized controlled trial, J Clin Oncol., 23(24):5560–5567, 2005.

37. S.M. Bentzen: Dose painting and theragnostic imaging: towards the prescription, planning and delivery of biologically targeted dose distributions in external beam radiation oncology, Cancer Treat Res., 139:41–62, 2008.

38. M. Baumann, C. Petersen, M. Krause: TCP and NTCP in preclinical and clinical research in Europe, Rays, 30(2):121–126, 2005.

39. A.J. Chalmers, S.M. Bentzen, F.M. Buffa: A general framework for quantifying the effects of DNA repair inhibitors on radiation sensitivity as a function of dose, Theor Biol Med Model. 4:25, 2007.

40. G.A. Plataniotis, R.G. Dale: Use of concept of chemotherapy-equivalent biologically effective dose to provide quantitative evaluation of contribution of chemotherapy to local tumor control in chemoradiotherapy cervical cancer trials, Int J Radiat Oncol Biol Phys., 72(5):1538–1543, 2008.

41. S.M. Bentzen, P.M. Harari, J. Bernier: Exploitable mechanisms for combining drugs with radiation: concepts, achievements and future directions, Nat Clin Pract Oncol., 4(3):172–180, 2007.

ACCURACY STRATEGIES IN RADIOTHERAPY

GÜNTHER H. HARTMANN
*German Cancer Research Center, Department of Medical
Physics in Radiation Oncology, Im Neuenheimer Feld 280,
D-69120 Heidelberg, Germany
g.hartmann@dkfz-heidelberg.de*

Abstract Radiotherapy is one of the most effective modalities for the treatment of cancer. Is it possible to increase its effectiveness? It is well known that there is a high degree of uncertainty associated with the target volume of most cancer sites. The sources of these uncertainties include, but are not limited to, the uncertainty involved in the delivery of the therapeutical dose, the delineation of the target volume, patient setup errors, the motion of the target, and patient movements. Such uncertainties will always influence the accuracy of radiotherapy. It is assumed that this directly has an impact on the effectiveness of radiotherapy. In order to develop strategies to improve the accuracy and hence effectiveness of radiotherapy it is therefore first needed to assess required and achievable uncertainties of all steps involved. Only based on such data strategies for improvements can be developed.

Keywords: Radiotherapy; accuracy; uncertainty; dosimetry

1. Introduction

1.1. CANCER INCIDENCE AND ROLE OF RADIOTHERAPY

The International Atomic Energy Agency (IAEA) has very recently issued a document on the design and implementation of radiotherapy programs to be applied throughout the world.[1] It represents in particular a response to the situation of increasing cancer diseases.

According to recent estimates of the International Agency for Research on Cancer (IARC) and the World Health Organization (WHO), approximately ten million new cancer cases are being detected per year world-wide, with slightly more than half of the cases occurring in developing countries.

Y. Lemoigne and A. Caner (eds.), *Radiotherapy and Brachytherapy*,
© Springer Science + Business Media B.V. 2009

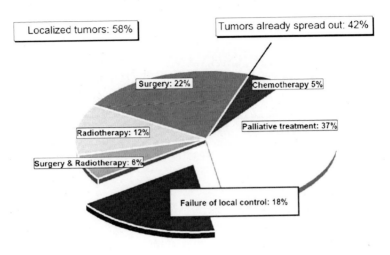

Figure 1. Current status of tumor treatments (EU, 1991).

By the year 2015 this number is expected to increase to about 15 million cases, of which two thirds will occur in developing countries. Treatment of cancer patients is therefore an increasing challenge also and even in particular in developing countries.

Radiotherapy can take over a significant role in the treatment of cancer as shown in Fig. 1: About half of all cancer patients receive radiotherapy, either as part of their primary treatment or in connection with recurrences or palliation. For Europe it was found that radiotherapy (alone/in combination) is successful in 18%. However, Fig. 1 also shows that within the group of still localized tumors (which should be principally curable by locally effective treatment modalities) the failure rate of local tumor control amounts to 18%. This is a real motivation to develop treatment strategies for further improvements. A possible strategy is the improvement of the accuracy in the dose delivery to the target volume.

In order to discuss possible strategies in a meaningful way, however, it is needed to take a close look into the problem: What degree of accuracy is required and can be really achieved?

1.2. ACCURACY AND UNCERTAINTY

Everybody has an idea of what accuracy means. However, although frequently used the term accuracy is only a qualitative expression. To be more quantitative, one needs an alternative concept.

Admittedly, there are indeed quantitative ideas behind the terms "Accuracy" and "Precision". These terms are frequently used to describe the quality of a measurement (or procedure) such as: "Accuracy" specifies the proximity of the mean value of a measurement to the true value. "Precision" specifies the degree of reproducibility of a measurement. The meaning and application of the terms of "Accuracy" and "Precision" is also illustrated in Figs. 2, 3.

The main disadvantage of these definitions is that it requires the knowledge of the so-called "true" value. The problem involved is: Does the "true" value really exist? The problem behind is illustrated in Fig. 4 by an example of measuring the thickness of a thin sheet of aluminum at a specified temperature.

First the material is brought close to the specified temperature and its thickness at a particular place is measured with a micrometer. Some corrections (temperature, temperature dependent calibration of the micrometer, pressure of the micrometer) must be then applied. The corrected result may be considered as the "true" value.

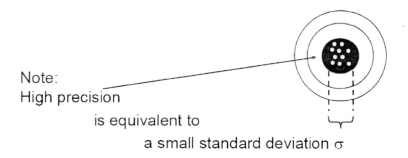

Figure 2. Relation between precision and standard deviation.

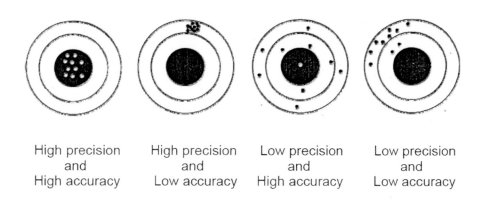

| High precision and High accuracy | High precision and Low accuracy | Low precision and High accuracy | Low precision and Low accuracy |

Figure 3. Examples for use of the terms "precision" and "accuracy".

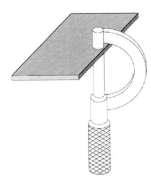

Figure 4. Example of a measurement of the thickness of a sheet of aluminum. It demonstrates that a very accurate value can certainly be obtained. However, this value may depend on the place were the micrometer is applied. The so-called "true" value therefore may ultimately remain unknown.

However: Had the micrometer been applied to a different place, perhaps another "true" value would have been obtained. Therefore the question arises: How to get the real "true value"? Answer: The "true value" probably exists, however it ultimately remains unknown! As a consequence, if the true value remains unknown, the value of the "error" and also that of "accuracy" can not be evaluated and must therefore also remain unknown.

This problem can be overcome by the introduction of concept of uncertainty as outlined in a document abbreviated as GUM.[2] It indeed provides an alternative concept because it defines uncertainty as a quantifiable attribute. It serves in particular as a clear procedure for characterizing the quality of a measurement (or quantitative statement). It is easily understood and generally accepted.

According to GUM, uncertainty is generally expressed by the quantitative parameter called the "standard uncertainty" of a result. "Standard uncertainty" is always calculated as the positive square root of the variance of the probability distribution (which is equivalent to the standard deviation). A combined (standard) uncertainty is obtained as the positive square root of the sum of variances weighted according to how the result is influenced by varying different influence components (if not correlated).

Influences that give rise to uncertainty are frequently differentiated as those of random nature or systematic nature. However, there is no principal difference between the uncertainty component arising from a random effect and that one arising from a correction for a systematic effect. All components that contribute to the uncertainty are of same nature and therefore are to be treated identically.

Type A uncertainty:	Type B uncertainty:
Those components which are evaluated by statistical analysis of series of observations	Those components which are evaluated by other means

Figure 5. The two methods for the evaluation of uncertainties: Type A and type B uncertainties.

The treatment of choice for all components is the use of an adequate probability distribution. For this purpose, all components to the total uncertainty are grouped into two different categories according to how the probability distribution is determined as shown in Fig. 5.

1.2.1. *Example of type A uncertainty evaluation*

Type A uncertainty evaluation refers to the situation where the uncertainty of a quantity can be obtained by performing repeated measurements. As an example, in a dosimetric measurement a charge Q is measured 20 times. Results have been grouped into charge intervals of 0.5 nC as shown in Fig. 6 left as a histogram of frequency of these charge intervals. The "best estimate" result is obtained according to equation (1) as the mean value of the single measurements of charge and shown in Fig. 6 right as red vertical dotted line.

$$\overline{Q} = \frac{1}{N}\sum_{i=1}^{N} Q_i \; z \tag{1}$$

The standard uncertainty of the measured charge is obtained as the estimate of the standard deviation of the mean charge and illustrated as orange area.

$$u_{\overline{Q}} = \sqrt{\frac{1}{N-1}\sum_{i=1}^{N}\left(Q_i - \overline{Q}\right)^2} \tag{2}$$

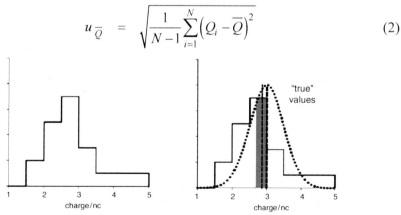

Figure 6. Type A uncertainty evaluation applied to repeated measurements of a charge. Charge values are grouped into charge intervals and shown as histogram (left). "Best estimate" value and range of standard uncertainty are shown in the right diagram as red dotted line and orange area "True values" (i.e. the really underlying distribution of charge) are also shown.

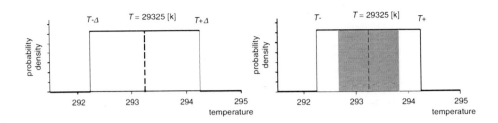

Figure 7. Type B uncertainty evaluation applied to the measurement of a temperature with limited information on the accuracy of the measurement.

1.2.2. *Example of type B uncertainty evaluation*

Type B uncertainty evaluation refers to a quantity of which the value is brought into the measurement from external sources like calibration factors or referenced data from handbooks. The corresponding distribution of possible values is evaluated based on all available information such as:

- Previous measurement
- Experience
- General knowledge
- Manufacturer's specification

Consider the case where a measured temperature T of 293.25 K is used as input quantity for the air density correction factor (Fig. 7). Frequently, little information is available on the accuracy of the temperature determination. All one can do is to suppose that there is a symmetric lower and upper bound around T (T − Δ, T + Δ), and that any value between this interval has an equal probability.

In order to calculate the best estimate and the standard deviation of the measured temperature, one has to establish the associated mathematical probability function p(T). It is a priori given by:

$$p(T) = \text{constant} \quad \text{for} \quad T - \Delta \leq x \leq T + \Delta$$
$$p(T) = 0 \quad \quad \text{otherwise} \tag{3}$$

Since the integral $\int_{T-\Delta}^{T+\Delta} p(T)dT$ must be equal to 1, it follows that p(T) is given by:

$$p(T) = 1/(2\Delta) \quad \text{for} \quad T - \Delta \leq x \leq T + \Delta$$
$$p(T) = 0 \quad \quad \text{otherwise} \tag{4}$$

Based on this probability function, one can obtain the best estimate and the standard deviation via the variance v exactly in the same way as done before:

$$\overline{T} = \int_{-\infty}^{+\infty} T \cdot p(T)dT = \frac{1}{2\Delta} \int_{T-\Delta}^{T+\Delta} T \, dT = T \tag{5}$$

$$v = \int_{-\infty}^{+\infty} (T-\overline{T})^2 \, p(T)dT = \frac{1}{2\Delta} \int_{T-\Delta}^{T+\Delta} (T-\overline{T})^2 \, dT = \frac{1}{3}\Delta^2 \tag{6}$$

$$u_{\overline{T}} = \sqrt{v} = \frac{\Delta}{\sqrt{3}} \tag{7}$$

In summary, the main idea of this concept is to treat any uncertainty (no matter whether random or systematic) by an appropriate mathematically defined probability function and the clear prescription of calculating the best estimate and its uncertainty by the mean and the positive square root of the variance of the mean. The combined (standard) uncertainty u_c of a quantity F which is dependent on a series of influence quantities x_i is obtained as the positive square root of the sum of variances weighted according to how the result is influenced by varying different influence components:

$$u_c = +\sqrt{\sum_{i=1}^{N} \left(\frac{\partial F}{\partial x_i}\right)^2 \cdot u^2(x_i)} \tag{8}$$

As a final remark to this concept, one should keep in mind that for the indication of the quality of a measurement, the term "error" should be avoided and substituted by the term "uncertainty". In the following sections this is generally applied.

2. Accuracy requirements from a clinical perspective

Many sources of uncertainty exist in radiotherapy procedures. They can be divided into two main categories: (1) the category related to dosimetry including issues such as beam calibration, relative measurements (data acquisition for treatment planning), or dose calculation in the treatment planning system, and (2) the category related to geometry including areas such as target delineation, setup, or organ motion.

2.1. DOSIMETRIC ASPECTS

2.1.1. *Requirements*

Quantitative date for limits of uncertainty (on tolerance or action levels) can be derived from experimental dose-effect curves or clinical observations. Mijnheer et al.[3] have derived a quantitative conclusion from clinical observations using the parameter $\Delta_{25,50}$. This parameter describes the relative dose increase in % giving raise of the probability for normal tissue reactions from 25% to 50%. Observations obtained in different studies are listed in Table 1. They found that on average this probability for most normal tissue reactions is 7%. Therefore, if the overall uncertainty in the absorbed dose is larger than 7%, any transfer of information from one center to another will introduce unacceptable risks on complications. It is assumed that this statement was derived for a confidence level of 95%. The assessed 7% uncertainty of dose must therefore be associated to a combined uncertainty being twice the standard uncertainty, i.e. the required standard combined uncertainty in the absorbed dose delivery will amount to 3.5%.

TABLE 1. The parameter $\Delta_{25,50}$.obtained from different clinical studies.

Normal tissue reaction	$\Delta_{25,50}$ (%)
Major chronic complications of the larynx	2
Peripheral neuropathy	3
Late skin damage	4
Late intestinal damage	4
Brachial plexus	5
Radiation pneumonitis	6
Skin reaction	7
Major complications of the intestine and bladder	9
Skin and lip	10
Myelitis	15
Major and not-major complications of the larynx	17

Further general guidelines derived for the required uncertainty in dose delivery are based on an approach described in a supplement in ACTA ONCOLOGICA edited by Anders Brahme.[4] Following this concept, the steepness of a dose effect curve $P(D)$ can be characterized directly by the relative gradient of the dose-effect curve called γ:

$$\gamma = \left(\frac{\Delta P}{\Delta D}\right) \cdot D \qquad (9)$$

This parameter is a dimensionless number which describes how large a change in tumor control probability is to be expected for a given relative increase in absorbed dose. Clinically observed values for γ at the steepest part of the dose response curve vary for normal tissues from 0.8 and 5. The range for tumors is between 0.4 and 9. In a good approximation one obtains a mean γ values between 3 and 5 for both.

Two different uncertainties of dose may occur:

- The variance in mean dose $\sigma_{\overline{D}}$ which reduces the steepness of the response curve
- The variation in dose around the mean dose σ_D which reduces the value of tumor control

Using the mathematical formalism as outlined in the appendix of supplement, one can derive required values of dose uncertainties at the tolerance and action level. Based on values of $\gamma = 3$ and $\gamma = 5$ one obtains relative standard deviations in dose delivery as shown in Table 2.

TABLE 2. Relative standard deviations in % in dose delivery at the tolerance and action level.

	Tolerance level $\sigma_P \leq 5\%$ and $\Delta P \geq 3\%$		Action level $\sigma_P \leq 10\%$ and $\Delta P \geq 5\%$	
	$\gamma = 3$	$\gamma = 5$	$\gamma = 3$	$\gamma = 5$
$\sigma_{\overline{D}}/\overline{D}$	2.5	1.5	3.5	2.0
$\sigma_D/\overline{D}$	4.5	3.0	7.0	4.5
$\sigma_{tot}/\overline{D}$	5.0	3.5	8.0	5.0

2.1.2. Assessed uncertainties in beam delivery

There are several sources of uncertainties in beam delivery. Most important are the uncertainties in beam calibration (i.e. the determination of absorbed dose per monitor units under reference conditions) and the uncertainties involved in the physical treatment planning.

The achievable uncertainty in beam calibration is well described in the IAEA dosimetry protocol TRS 398.[5] When an ionization chamber is used for the calibration in a beam quality Q that differs from the quality Q_0 used in the chamber calibration at the standards laboratory, the absorbed dose to water is:

$$D_{w,Q_0} = M_{Q_0} \, N_{D,w,Q_0} \cdot k_{Q,Q_0} \quad (10)$$

M_{Q_o} is the reading of the dosimeter under the reference conditions used in the standards laboratory and corrected for influence quantities

N_{D,w,Q_o} is the calibration factor in terms of absorbed dose to water of the dosimeter obtained from a standards laboratory

k_{Q,Q_o} is now a factor correcting for the differences between the reference beam quality Qo and the actual user quality Q

The associated uncertainties in all steps of the calibration procedure are listed in Table 3. The achievable value of uncertainty amounts to 1.5%.

TABLE 3. Estimated relative standard uncertainty a of dw,q at the reference depth in water and for a high-energy photon beam, based on a chamber calibration in ^{60}Co Gamma radiation.

Physical quantity or procedure	Rel. stand. uncertainty (%)
Step 1: Standards Laboratory	
$N_{D,w}$ calibration of secondary standard at PSDL	0.5
Long term stability of secondary standard	0.1
$N_{D,w}$ calibration of the user dosimeter at the standard laboratory	0.4
Combined uncertainty of Step 1	0.6
Step 2: User high-energy photon beam	
Long-term stability of user dosimeter	0.3
Establishment of reference conditions	0.4
Dosimeter reading M_Q relative to beam monitor	0.6
Correction for influence quantities k_i	0.4
Beam quality correction k_Q (calculated values)	1.0 [c]
Combined uncertainty of Step 2	1.4
Combined standard uncertainty of $D_{w,Q}$ (Steps 1 + 2)	1.5

The dosimetric accuracy achievable for treatment planning purposes has been the subject of much discussion. Here a method for characterization of the accuracy of a dose calculation method similar to that used by Van Dyk et al.[6] is used. For this purpose, the calculation of dose distribution due to a beam is broken up into several regions, as illustrated in Fig. 8. The typical uncertainties associated to the different regions are also shown.

Although some years ago, there are also estimates on the total standard uncertainty in beam delivery carried out by Svensson (1983) and Mijnheer (1987). Various sources of uncertainty have been taken into account for that. Data are summarized in Table 4.

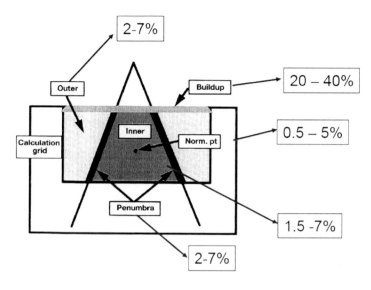

Figure 8. Regions for uncertainties photon dose calculation. (Picture taken from Fraas et al.[7])

TABLE 4. Standard and total standard uncertainty in beam delivery in %.

	Svensson (1983)	Mijnheer (1987)
Calibration of dosemeter	2.1	1.4
Dose at reference point	4.5	2.1
Dose distribution	3.0	1.4
Dose in patient	4.4	2.1
Patient/beam setup	4.0	2.1
Total uncertainty	**8.3**	**4.4**

2.2. ANATOMICAL/GEOMETRICAL ASPECTS

There are even more sources of uncertainties involved in the category related to the anatomy and geometry of the patient. Most important are the uncertainties in delineation of the target volume (and in particular of the clinical target volume), in the patient setup, and in movements of the target volume at both types, inter-fractional and intra-fractional. While it is impossible to eliminate all these uncer-tainties, an understanding of the source and magnitude of these errors is essential to reduce them.

2.2.1. *Target delineation*

Many studies on the variation in the determination of the target volume have been published. Only three examples are given in the following sections.

Kagawa et al have performed a study with 22 patients in 1997 on the localization of the prostate for 3D conformal radiation therapy using CT-MRI image fusion software.[8] Differences between CT and MRI contours were assessed by one radiation oncologists. A typical result is shown in Fig. 9. The average ratio between CT and MRI volume was 1.24.

Rasch et al. have investigated the potential impact of CT-MRI matching on tumor volume delineation in advanced head and neck cancer.[9] The gross tumor volume was contoured at six patients by four observers (Fig. 10). The average ratio between CT and MRI GTV was 1.3. CT and MRI volumes were found to be complementary, i.e. neither volume encompassed the other. MRI had less variability.

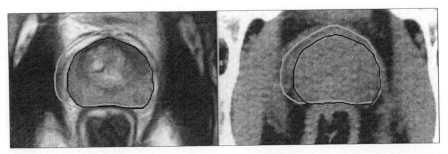

Figure 9. A view of CT-MRI image fusion of the prostate. The prostate contour is outlined on CT (right: grey line) and on corresponding reconstructed T2-weighted MRI (left: black line).

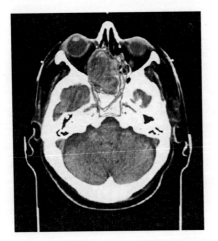 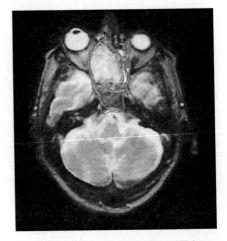

Figure 10. Four contours as outlined by four observers; contours in red outlined on CT image (right); contours in green outlined on axial MRI image (left). See color picture in Appendix I.

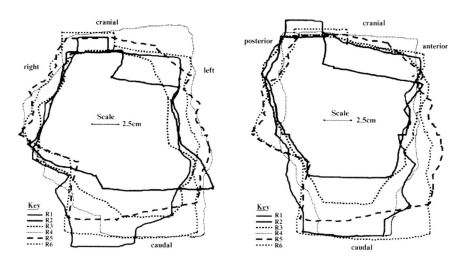

Figure 11. Beams eye view of manually contoured PTV's; left: frontal plane; right: sagittal plane. (Pictures taken from Senan et al., Radiother Oncol 53, 1999.)

Senan et al. have evaluated the target contouring in non-small cell lung cancer.[10] After implementation of a delineation protocol inter-observer variations were determined in three patients by six radiation oncologists (Fig. 11). A ratio of two was found between the largest and the smallest PTV.

2.2.2. *Setup uncertainties*

Uncertainties involved in the setup procedure may be divided into two categories: (1) those due to treatment execution variations (also called random or day-today variations) and (2) those due to treatment preparation variations. The latter errors are often called systematic, because they are systematic for a single radiotherapy course of a single patient, but they are stochastic over a group of patients. Van Kerk et al.[11] have performed an illustrative simulation of all the uncertainties involved in the treatment of prostate cancer for deriving treatment margins. The visualization of the effect of tumor motion, setup error, and delineation uncertainties on the dose delivery in radiotherapy is shown in Fig. 12. Typical values of setup uncertainties are listed in Table 5.

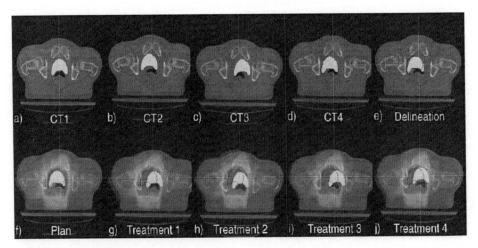

Figure 12. Simulated visualization of the effect of uncertainties on the dose delivery in radiotherapy. (a) Assume that the indicated region is the "true" clinical target volume (CTV), i.e., the region with tumor cells. The (white) cross wires indicate the room lasers. (b–d) Due to setup error on the CT scanner and organ motion, the four CT scans are not identical. (e) CT4 is used for planning, and the delineation (black contour) adds some extra error, because the "true" CTV is invisible. (f) The planning is based on the delineated CTV. (g–i) For each treatment fraction, the error made in the treatment preparation phase is reproduced, causing a systematic shift of the "true" CTV relative to the delivered dose distribution. In addition, random movements (treatment execution variations) occur due to tumor motion and setup error. (Picture taken from van Herk et al.[11]) See color picture in Appendix I.

TABLE 5. Setup uncertainties in mm.

	Unimmobilized (mm)	Immobilized (mm)
Abdomen and Pelvis	6–8	3–4
Breast		3
Thorax	4–6	
Head and Neck	10	3–4
Intracranial Targets	3	mask: 2
		head frame: <1

2.2.3. *Organ motion*

Variations in patient position and movement can be minimized with the help of precise patient positioning systems and rigid immobilization devices. For some anatomical sites, however, the internal motion of organs due to physiological processes presents a further source of uncertainty and hence a challenge for improvements. Langen and Jones have presented a valuable review on available data.[12] As an example, Table 6 is showing the inter-fraction motion for gyneco-logical tumors, for the bladder and for the rectum. A summary of geometrical uncertainties is given in Table 7.

TABLE 6. Inter-fraction organ motions. (Data taken from Langen and Jones.[12])

Organ	Parameter	Movement/change
Gynecological tumors	Corpus uteri:	Mean SI movement = 7 mm
	Cervix mean:	SI movement = 4 mm
Bladder	Bladder wall displacement:	Up to 27 mm
	Frequent wall displacement:	>15 mm
	volume changes:	Up to 40% (1 SD)
Rectum	Volume changes:	Up to 28% (1 SD)
	Diameter change:	Up to 46 mm (1 SD)

TABLE 7. Summary of geometrical uncertainties.

Site	Delineation	Set-up	Organ motion
Prostate	3–4 mm	3–4 mm	3–4 mm
Bladder	1.5–3 mm	3–4 mm	10–20 mm
Lung	up to 50 mm	4–6 mm	0–20 mm
H + N	factor 5 CTV	2 mm	rigid
Breast	6–42 mm	4–6 mm	6 mm

3. Possible radiotherapy strategies and summary

It has already initially pointed out that possible strategies for improvements in radiotherapy based on the assumption that a higher accuracy will lead to a higher effectiveness of radiotherapy require a thorough study of the following problem: Which degree of accuracy should be aimed at and which degree can be really achieved in practice? Many aspects are involved as frequently demonstrated by the so-called "chain of radiotherapy" as shown in Fig. 13. From that it is immediately obvious that only if each aspect is adequately taken into account, a substantial improvement can be expected.

This contribution does not aim to offer final conclusion. Nevertheless, for the dosimetrical aspects some very rough ideas are outlined next.

The ICRU Report 24 from 1976[13] claims that there is evidence (for certain types of tumor) that points to the need for an accuracy of ±5% in the delivery of an absorbed dose to a target volume if the eradication of the primary tumor is

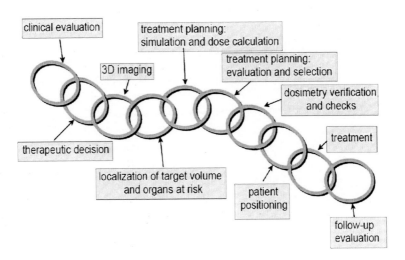

Figure 13. The "chain of radiotherapy".

sought. This statement is given in a context where uncertainties were estimated at the 95% confidence level, and have been interpreted as if they correspond to approximately two standard deviations. Thus the requirement corresponds to a combined uncertainty of 2.5% at the level of one standard deviation. Mijnheer et al.[3] proposed a requirement on accuracy of 3.5%.

Today it is considered that a goal in dose delivery to the patient based on such an accuracy requirement is too strict and the figure should be increased to about one standard deviation of 5%, but there are still no definite recommendations in this respect. This recommendation for an accuracy of ±5% could be also interpreted as a tolerance for the deviation between the prescribed dose and the dose delivered to the target volume. It appears that such recommendations have a realistic chance to fulfilled in practice, however, further improvement will strongly depend in particular on the application of well designed quality assurance programs and on innovations in both areas, procedures and equipment.

References

1. IAEA: Setting up a Radiotherapy Programme: Clinical, Medical Physics, Radiation Protection and Safety Aspects. STI/PUB/1296, Vienna, International Atomic Energy Agency, 2008 [www-pub.iaea.org/MTCD/publications/PDF/pub1296_web.pdf].
2. International Organization for Standardization: "Guide to the Expression of Uncertainty in Measurement", 2nd edn, Geneva (1995).

3. Mijnheer BJ, Battermann JJ, Wambersie A: What degree of accuracy is required and can be achieved in photon and neutron therapy? Radiother Oncol 8, 237–252 (1987).
4. ACTA ONCOLOGICA, Suppl.1 (edited by Brahme A), Stockholm (1988).
5. IAEA International Atomic Energy Agency. "Absorbed Dose Determination in External Beam Radiotherapy: An International Code of Practice for Dosimetry based on Standards of Absorbed Dose to Water". Technical Report Series no. 398, IAEA, Vienna (2000).
6. Van Dyk J, Barnett R, Cygler J, Shragge P: Commissioning and quality assurance of treatment planning computers. Int J Radiat Oncol Biol Phys 26, 261–273 (1993).
7. Fraass B, Doppke K, Hunt M, Kutcher G, Starkschall G, Stern R, Van Dyke J: AAPM Radiation Therapy Committee Task Group 53: Quality assurance for clinical radiotherapy treatment planning. Med Phys 25, 1773–1829 (1998).
8. Kagawa K, Lee WR, Schultheiss TE, Hunt MA, Shaer AH, Hanks GE: Initial clinical assessment of CT-MRI image fusion software in localization of the prostate for 3D conformal radiation therapy. Int J Radiat Oncol Biol Phys 38, 319–325 (1997).
9. Rasch C, Keus R, Pameijer FA, Koops W, de Ru V, Muller S, Touw A, Bartelink H, van Herk M, Lebesque JV: The potential impact of CT-MRI matching on tumor volume delineation in advanced head and neck cancer. Int J Radiat Oncol Biol Phys 39, 841–848 (1997).
10. Senan S, van Sörnsen de Koste J, Samson M, Tankink H, Jansen P, Nowak PJ, Krol AD, Schmitz P, Lagerwaard FJ: Evaluation of a target contouring protocol for 3D conformal radiotherapy in non-small cell lung cancer. Radiother Oncol. 53, 247–255 (1999).
11. van Herk M, Remeijer P, Rasch C, Lebesque JV: The probability of correct target dosage: dose-population histograms for deriving treatment margins in radiotherapy. Int J Radiat Oncol Biol Phys 47, 1121–1135 (2000).
12. Langen KM, Jones DT: Organ motion and its management. Int J Radiat Oncol Biol Phys 50, 265–278 (2001).
13. ICRU International Commission on Radiation. Determination of absorbed dose in a patient irradiated by beams of X or gamma rays in radiotherapy procedures. ICRU Report 24, ICRU, Bethesda (1976).

PART II:

X-RAYS TELETHERAPY

DOSIMETRY FOR PHOTON AND ELECTRON RADIATION

GÜNTHER H. HARTMANN
*German Cancer Research Center, Department of Medical
Physics in Radiation Oncology, Im Neuenheimer Feld 280,
D-69120 Heidelberg, Germany
g.hartmann@dkfz-heidelberg.de*

Abstract The dose of radiation is a central issue in radiotherapy. Hence dosimetry must be well understood for anyone working in this field. In this chapter theoretical aspects as well as practical aspects of the dosimetry of high energy photons and electrons are addressed. The practical aspects are exclusively dedicated to the determination of absorbed dose to water in terms of Gray using ionization chambers under so-called reference conditions frequently referred to as beam calibration. The procedures described for that are following the concepts as outlined in the IAEA document "Absorbed Dose Determination in External Beam Radiotherapy: An International Code of Practice for Dosimetry based on Standards of Absorbed Dose to Water", Technical Report Series no. 398.

Keywords: Absorbed dose; radiation detectors; ionization chamber; dosimetry protocol; beam calibration; cross calibration

1.   Introduction: definition of "radiation dose"

1.1.   EXACT PHYSICAL MEANING OF THE TERM "DOSE"

"Dose" is a somewhat sloppy expression to denote the dose of radiation and should be used only if the partner of communication really knows what you are meaning. A dose of radiation is correctly denoted by the physical quantity of absorbed dose. It may be interesting that the use of the term "dose of radiation" is derived from the dose of medicine given to a patient to cure him.

Figure 1. A "dose" of medicine given to a patient. (Picture taken from the famous children's book "Der Struwwelpeter".)

The most fundamental definition of the quantity absorbed dose is given in ICRU 60.[1] According to ICRU 60, the absorbed dose D is defined by:

$$D = \frac{d\bar{\varepsilon}}{dm} \tag{1}$$

where:

$d\bar{\varepsilon}$ is the mean energy imparted to matter of mass (see Fig. 2)

dm is a small (infinitesimal) element of mass

The unit of absorbed dose is joule per kilogram (J/kg), the special name for this unit is gray (Gy).

1.2. CHARACTERISTICS OF ABSORBED DOSE

The term "energy imparted" as used in the definition above is the radiation energy absorbed in a volume V. Therefore, the term "absorbed dose" refers to an exactly defined volume and only to the volume V. Furthermore, the term "absorbed dose" refers to the material contained in this volume as also illustrated in Fig. 2.

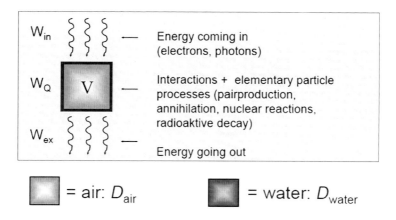

Figure 2. Absorbed dose being the energy imparted to matter within a volume V.

"Absorbed dose" is a macroscopic quantity that refers to a point in space:

$$D = D\left(\vec{r}\right) \tag{2}$$

This is associated with: (a) the quantity of D is steadily in space and time at $\vec{r}$, and (b) D can be differentiated in space and time. Is this a contradiction to its characteristic to refer to an exactly defined volume and not to a point in space?

The answer is implicitly contained in the basic definition of equation (1). The **energy imparted** to matter in a given volume, ε, is the sum of all energy deposits in a defined volume, i.e. the energy deposits by all those basic inter-action processes which have occurred in the volume during a time interval considered:

$$\varepsilon = \sum_i \varepsilon_i \tag{3}$$

Referring to Fig. 3 which is a schematic illustration of energy deposits occurring with photons, energy imparted is the energy deposited only within the volume V along the tracks of the secondary electrons set in motion by the interaction (red points) of the photons which, however, may occur inside and outside the volume V. An example for a single energy deposit ε_i is the positron annihilation event as shown in Fig. 4.

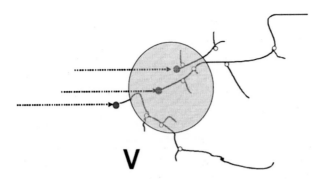

Figure 3. Energy depositions involved for the definition of absorbed dose: only those deposits within the volume V along the tracks of secondary electrons emitted by interactions of photons (grey points).

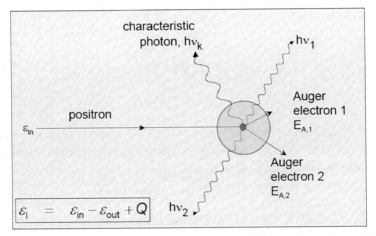

Figure 4. A single energy deposit occurring from the annihilation of a positron. Please note: $\varepsilon_{out} = h\nu_1 + h\nu_2 + h\nu_k + E_{A,1} + E_{A,2}$ and $Q = 2m_0c^2$.

The term "**energy deposit**" refers to a single interaction process, i.e. it is the energy deposited in a single interaction i thus that:

$$\varepsilon_i = \varepsilon_{in} - \varepsilon_{out} + Q \qquad \text{Unit: J} \qquad (4)$$

where:

ε_{in} is the energy of the incident ionizing particle (excluding rest energy)

ε_{out} is the sum of energies of all ionizing particles leaving the interaction (excluding rest energy)

Q is the change in the rest energies of the nucleus and of all particles involved in the interaction

It follows from the character of an energy deposit that the energy imparted ε is also a stochastic quantity. That means that if the determination of ε is repeated, it will never will yield the same value. Instead of, it will be subjected to a probability distribution. The spread of the distribution will be larger and larger with a decreasing number of energy deposits. This may happen if the size of the volume (for which the dose is defined) is approaching the infinitesimal value dm. It also may happen for low dose rates as normally encountered in the context of radiation protection. For a large volume or mass element the distribution gets very narrow approaching the value of absorbed dose according to the definition of (1) as illustrated in Fig. 5.

In order to avoid that the quantity absorbed dose is subjected to the stochastic character of the energy transfer from radiation to matter, it was defined in equation (1) using the **mean energy imparted** and not simply the energy imparted itself. From that it follows that the absorbed dose D is a non-stochastic quantity which is indeed steadily in space and time, and which can be differentiated in space and time.

It should be mentioned however, that the stochastic character of energy transfer may have an impact at a microscopic level, i.e. at the dimension of a single cell and smaller. The pattern of the distribution of energy deposits at the microscopic level depends in particularly on the type of radiation. The pattern might be quite different for low LET and high LET radiation. Therefore, in the field of radiobiology it may be possible that the specific pattern of energy distribution must be taken into account in order to understand a certain phenomena.

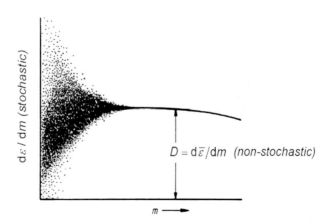

Figure 5. Change of the distribution of the stochastic quantity dε/dm depending on the mass m of dm.

2. General methods of dose measurement

Absorbed dose is measured with a (radiation) dosimeter. The four most commonly used types of radiation dosimeters are:

Ionization chamber
Radiographic or radiochromic films
Thermo-luminescence dosimeter (TLD)
Diode

Examples of different ionization chambers are shown in Fig. 6.

Advantage and disadvantage of these radiation dosimeters are summarized in the following Tables 1–4. It should be added that for radiochromic films two disadvantages as listed in Table 2 do not apply: (a) processing facilities and a darkroom are not required any more, and (b) the problems with the energy dependence are significantly smaller.

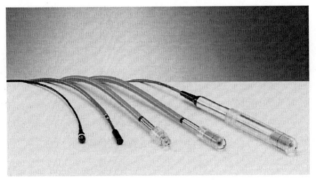

Figure 6. Different ionization chambers as used in radiotherapy. (Picture taken from PTW.)

TABLE 1. Properties of ionization chambers.

Advantage	Disadvantage
Accurate and precise	Connecting cables required
Recommended for beam calibration	High voltage supply required
Necessary corrections well understood	Many corrections required
Instant readout	

TABLE 2. Properties of radiographic films.

Advantage	Disadvantage
2-D spatial resolution	Darkroom and processing facilities required
Very thin: does not perturb the beam	Processing difficult to control
	Variation between films & batches
	Needs proper calibration against ionization chambers
	Energy dependence problems
	Cannot be used for beam calibration

TABLE 3. Properties of thermo-luminescence dosimeter.

Advantage	Disadvantage
Small in size: point dose measurements possible	Signal erased during readout
	Easy to lose reading
Many TLDs can be exposed in a single exposure	No instant readout
Available in various forms	Accurate results require care
Some are reasonably tissue equivalent	Readout and calibration time consuming
Not expensive	Not recommended for beam calibration

TABLE 4. Properties of diodes.

Advantage	Disadvantage
Small size	Requires connecting cables
High sensitivity	Variability of calibration with temperature
Instant readout	Change in sensitivity with accumulated dose
No external bias voltage	Special care needed to ensure constancy of response
Simple instrumentation	Cannot be used for beam calibration

The following sections are exclusively dedicated to the determination of absorbed dose to water (in terms of Gray) using ionization chambers. In doing that, we are strictly following the concepts and procedures as outlined in the IAEA document "Absorbed Dose Determination in External Beam Radio-therapy: An International Code of Practice for Dosimetry based on Standards of Absorbed Dose to Water", Technical Report Series no. 398.[2]

3.   Principles of dosimetry with ionization chambers

Measurement of absorbed dose requires the measurement of the mean energy imparted in small volume, i.e. the energy transferred into this volume by the various interaction processes. The result from such interaction processes normally is the creation of ion pairs. The creation and measurement of ionization in a gas is the basis for dosimetry with ionization chambers. Because of the key role that ionization chambers play in radiotherapy dosimetry, it is vital that practicing physicists have a thorough knowledge of their characteristics.

The ionization chamber is the most practical and most widely used type of dosimeter for accurate measurement of machine output in radiotherapy. It may be used as an absolute or relative dosimeter. Its sensitive volume is usually filled with ambient air.

3.1. CHARGE MEASUREMENT

The measured quantity which is related to dose is the charge produced in the sensitive volume of the ionization chamber. The measured quantity which is related to the dose rate is the current.

Measured charge Q and sensitive air mass m_{air} are related to absorbed dose in air D_{air} by:

$$D_{air} = \frac{Q}{m_{air}} \left(\frac{\overline{W}_{air}}{e} \right) \tag{5}$$

$\left(\overline{W}_{air} / e \right)$ is the mean energy required to produce an ion pair in air per unit charge e. It is generally assumed that a constant value can be used, valid for the complete photon and electron energy range used in radiotherapy dosimetry. It depends on relative humidity of air such that

$(\overline{W}_{air}/e) = 33.77$ J/C for air at relative humidity of 50%:

$(\overline{W}_{air}/e) = 33.97$ J/C for dry air:

3.2. BRAGG-GRAY (BG) CONDITION

Following this, the measured absorbed dose in air D_{air} must be converted into absorbed dose in water D_w. This conversion depends on several conditions such as type and energy of radiation or type and volume of the ionization chamber. For most cases in clinically applied radiation fields such as for high energy photons (E > 1 MeV) or high energy electrons, the so-called Bragg-Gray cavity theory can be applied. This theory can be applied if the two following conditions are fulfilled.

Condition 1: The cavity must be small when compared with the range of charged particles incident on it, so that its presence does not perturb the fluence of charged secondary electrons in the medium.

Consequences of the Bragg-Gray conditions 1: This condition is valid only in regions of charged particle equilibrium or transient charged particle equilibrium. However, the presence of a cavity always causes some degree of fluence perturbation. Therefore the introduction of a fluence perturbation correction factor is required if the cavity theory should be applied.

Condition 2: The absorbed dose in the cavity is deposited solely by charged particles crossing the cavity. Such electrons are also called "crossers".

Consequences of the Bragg-Gray conditions 2: Photon interactions in the cavity are assumed negligible and thus any possibly associated energy deposit is ignored. All electrons depositing energy inside the cavity are produced outside the cavity and completely cross the cavity. No secondary electrons are produced inside the cavity (starters) and no electron stops within the cavity (stoppers).

3.3. ABSORBED DOSE IN WATER

If Bragg-Gray conditions strictly apply, the measured absorbed dose in air D_{air} can be converted into absorbed dose in water D_w by:

$$D_w = D_{air} \cdot s_{w,air} \qquad (6)$$

where $s_{w,air}$ is the ratio water to air of the mean mass stopping power of the secondary electrons crossing the cavity.

If Bragg-Gray conditions apply only in a good approximation, the measured absorbed dose in air D_{air} can be converted into absorbed dose in water D_w by:

$$D_w = D_{air} \cdot s_{w,air}^{SA} \cdot p \qquad (7)$$

where $s_{w,air}^{SA}$ is the ratio water to air of the mean Spencer-Attix mass stopping power and p is an additional correction factor being the product for all per-turbation correction factors required to take into account deviations from BG-conditions. The calculation of Spencer-Attix stopping powers is based on a more realistic description of energy deposition of the secondary electrons including the stoppers. For more details the reader is referred to relevant textbooks.

4. Equipment for ionometric measurements

4.1. IONIZATION CHAMBERS

There are essentially two types of chambers as illustrated in Fig. 7:
   Cylindrical (also called thimble) chambers which are used for the calibration of:

(Orthovoltage x-ray beams)
Megavoltage x-ray beams
Electron beams with energies of 10 MeV and above

   Parallel-plate (also called end window or plane-parallel) chambers which are used for:

The calibration of superficial x-ray beams
The calibration of electron beams with energies below 10 MeV
Dose measurements in photon beams in the buildup region and surface dose

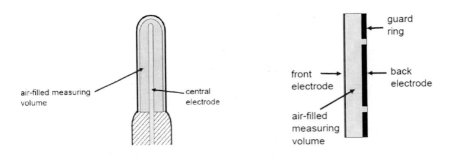

Figure 7. General design of cylindrical (left) and plane-parallel (right) chambers.

4.2. ELECTROMETER

An electrometer is a high gain, negative feedback, operational amplifier with a standard resistor or a standard capacitor in the feedback path to measure the chamber current and charge, respectively, collected over a fixed time interval (Fig. 8).

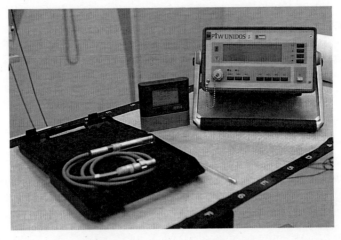

Figure 8. Assembly of electrometer and ionization chamber. (Picture taken from Chapter 3 of the IAEA Slides to Radiation Oncology Physics handbook, www- naweb.iaea.org/nahu/dmrp/slides .shtm.)

4.3. PHANTOMS

Phantom is a common name for materials that are used to replace the patient in studies of radiation interactions, in particular related to dosimetry in patients. Phantom material should meet the following criteria:

It absorbs photons in the same manner as tissue.
It scatters photons in the same manner as tissue.
It has the same density as tissue.
It contains the same number of electrons per gram as tissue.
It has the same effective atomic number as tissue.

Common plastic phantom materials used in dosimetry measurements are (Fig. 9):

Polystyrene (density: 0.96–1.04 g/cm^3)
Lucite (also called acrylic, plexiglass, polymethylmethacrylate, PMMA) with density of 1.18 g/cm^3
A-150 tissue equivalent plastic
Solid Water
Plastic water
Virtual water

With respect to beam calibration one should keep in mind: water is always recommended in the IAEA Codes of Practice as the most appropriate phantom material, for the calibration of megavoltage photon as well as for electron beams. The water phantom should extend to at least 5 cm beyond all four sides of the largest field size employed at the depth of measurement. There should also be a margin of at least 5 g/cm^2 beyond the maximum depth of measurement except for medium energy X rays in which case it should extend to at least 10 g/cm^2.

5. Principles of a calibration procedure

5.1. NEED FOR A PROTOCOL

Dosimetry protocols or codes of practice define and describe the procedures to be followed when calibrating a clinical photon or electron beam. The choice of which protocol to use is left to individual radiotherapy departments or jurisdictions. Dosimetry protocols are generally issued by national, regional, or international organizations.

Examples of dosimetry protocols:

on national level:
- UK: Institution of Physics and Engineering in Medicine and Biology (IPEMB)
- Germany: DIN 6800-2, Deutsches Institut für Normung

on regional level:
- American Association of Physicists in Medicine (AAPM) for North America: TG-51

- Nederlandse Commissie voor Stralingsdosimetrie (NCS) for Netherlands and Belgium
- Nordic Association of Clinical Physics (NACP) for Scandinavia

on international level:

- International Atomic Energy Agency (IAEA): Absorbed Dose Determination in External Beam Radiotherapy: An International Code of Practice for Dosimetry based on Standards of Absorbed Dose to Water.TRS 398

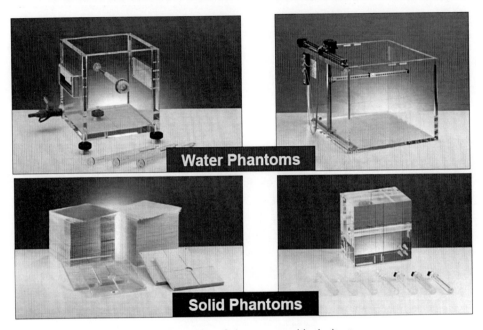

Figure 9. Examples of phantoms used in dosimetry.

A dosimetry protocol provides two essentials: (a) the formalism, and (b) all the data to be used to apply calibrated ionization chamber traceable to a standards laboratory for "dosimetry". "Dosimetry" in this context means the determination of absorbed dose to water under reference conditions in the clinical beam of a beam delivery unit.

Two types of dosimetry protocol are currently in use:

Protocols based on air kerma in air calibration coefficients
Protocols based on absorbed dose to water calibration coefficients

Conceptually, both types of protocol are similar and define the steps to be used in the process of determining absorbed dose from a signal measured by an ionization chamber. Here protocols based on absorbed dose to water calibration coefficients are treated only Table 5.

TABLE 5. Most frequently used reference conditions to calibrate ionization chambers.

Reference condition	Value
Beam quality	^{60}Co gamma radiation
Field size	10 x 10 cm
SSD	100 cm
Phantom	Water phantom
Measurement depth in water	5 cm
Positioning of chamber	Central electrode at measuring depth

5.2. CALIBRATION AND CALIBRATION COEFFICIENT

Suppose the absorbed dose is well known at a point in a water phantom under so-called reference conditions: the user chamber is then placed with its reference point at a depth of 5 cm in a water phantom and its calibration factor (or calibration coefficient) $N_{D,w}$ is obtained from:

$$N_{D,w,Co} = \frac{D_w}{M} \tag{8}$$

where M is the dosimeter reading. This normally is the measured charge. The symbol M is now used for that in the subsequent sections in order to be consistent with the symbols used in many dosimetry protocols.

The absorbed dose to water at the reference depth z_{ref} in water for a reference beam of quality Q_0 and in the absence of the chamber is given by

$$D_{w,Q_0} = M_{Q_0} N_{D,w,Q_0} \tag{9}$$

Where:

M_{Q_0} is the reading of the dosimeter under the reference conditions used in the standards laboratory and corrected for influence quantities.

N_{D,w,Q_0} is the calibration factor in terms of absorbed dose to water of the dosimeter obtained from a standards laboratory.

Subsequently, the chamber is to be used in a beam with another quality Q such as high energy photons or high energy electrons for which the quality of radiation differs from the calibration quality used at the standards laboratory. For this purpose the formula for the determination of absorbed dose to water must be changed to:

$$D_{w,Q_0} = M_{Q_0} N_{D,w,Q_0} \cdot k_{Q,Q_0} \tag{10}$$

where k_{Q,Q_0} is a further correction factor correcting for the differences between the reference beam quality Q_0 and the actual user quality Q.

Frequently, the common reference quality Q_0 used for the calibration of ionization chambers is the cobalt-60 gamma radiation in which case the shorter symbol k_Q is normally used to designate the beam quality correction factor:

How to get the beam quality correction factor? There are several alternatives:

1. An experimentally obtained is available.

    When no experimental data are available, or it is difficult to measure directly for realistic clinical beams, calculated correction factors must be used.

Values for k_Q are dependent on the quality of radiation (type, energy, machine). Therefore, each type of ionization chamber needs a particular value for k_Q. Those calculated values of k_Q are normally provided in the dosimetry protocols for a large variety of beam qualities and chambers (e.g. in TRS 398).

6. Performance of a calibration procedure

6.1. POSITIONING OF THE IONIZATION CHAMBER IN WATER

The absorbed dose to water is to be determined in a point P in water at the reference depth z_{ref}. This applies to the reference beam quality Q_0 as well as to the beam quality Q. According to the IAEA Dosimetry protocol the following depth for z_{ref} must be used:

for high energy photons	$z_{ref} = 10$ g cm^{-2} (or 5 g cm^{-2})	for $TPR_{20,10} < 0.7$
	$z_{ref} = 10$ g cm^{-2}	for $TPR_{20,10} \geq 0.7$
	where $TPR_{20,10}$ is the quality index of the high energy photons (see section 6.3.1)	
for high energy electrons	$z_{ref} = 0.6\, R_{50} - 0.1$ (in g cm^{-2})	
	where R_{50} is the quality index of the high energy electrons (see 6.3.2)	

Positioning can be practically performed by a prescription of how the reference point of a chamber must be adjusted in the water phantom in relation to z_{ref}.

For cylindrical chambers the reference point is at the centre of the cavity volume of the chamber on the chamber axis. For plane-parallel ionization chambers, the reference point is at the center of the front surface of the inner air cavity (Fig. 10).

In the absence of the chamber, the dose at a point P may be expressed by $D_{w,Q}(P)$. Using the chamber, the dose is determined by:

$$D_{w,Q}(P) = M_Q(P) \cdot N_{D,w} \cdot k_Q \qquad (11)$$

How the chamber must be positioned to measure that charge $M_Q(P)$ which correctly gives the dose at the point, $D_{w,Q}(P)$ using the basic formula (7)? The problem involved is: A chamber positioned with its cavity center at the point P does not sample the same electron fluence which is present at P in the undisturbed phantom, i.e. without the chamber as illustrated in Fig. 11.

The concept of the effective point of measurement, P_{eff} was introduced to handle this problem. It is that point in the chamber which when adjusted to the measuring depth will provide the dose in water using the basic equation (7). For cylindrical ionization chambers in standard calibration geometry, i.e. a radiation beam incident from one direction, P_{eff} is shifted from the position of the centre towards the source by a distance which depends on the type of beam and chamber. For plane-parallel ionization chambers P_{eff} is usually assumed to be situated in the center of the front surface of the air cavity.

One may therefore think that the correct way of positioning the chamber consists in the positioning at its effective point of measurement.

However: In the calibration procedure it does not really matter as long as the positioning is clearly defined (using a prescription of how the reference point of the chamber must be adjusted in relation to the measurement depth), and any

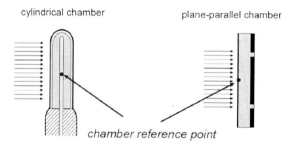

Figure 10. Reference point of an ionization chambers.

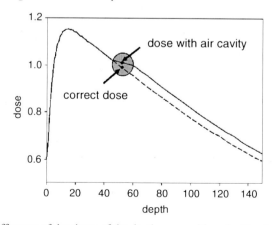

Figure 11. Difference of the shape of the depth curve with and without an air cavity.

deviation of the "correct" positioning (i.e. using P_{eff}) is taken into account in the calibration factor, or in the quality correction factor. It follows that if one is using the data of a particular dosimetry protocol, one has strictly apply the associated prescriptions of positioning (e.g. Tab. 6).

TABLE 6. Positioning of the reference point of a cylindrical chamber with inner radius r according to IAEA TRS 398.

	Purpose	
	Beam calibration	Depth dose measurement
Co-60	At measuring depth	0.6 r deeper than measuring depth
HE photons	At measuring depth	0.6 r deeper than measuring depth
HE electrons	0.5 r deeper than measuring depth	0.5 r deeper than measuring depth

6.2. DETERMINATION OF CHARGE UNDER REFERENCE CONDITIONS

The numerical value of the calibration factor and that of the quality correction factor are applicable only if the reference conditions are fulfilled. Reference conditions are described by a set of values of influence quantities. For a calibration in ^{60}C0 beam they are listed in Table 7.

TABLE 7. Complete set of reference conditions for calibration in a Co-60 beam.

Influence quantity	Reference value or reference characteristic
Phantom material	Water
Phantom size	30 x 30 x 30 cm (approximately)
Source-chamber distance	100 cm
Air temperature	$T_0 = 20°C$ c
Air pressure	$P_0 = 101.3$ kPa
Reference point of the ionization chamber	For cylindrical chambers, on the chamber axis
	For plane-parallel chambers on the inner surface of the entrance window
Depth in phantom of the reference point of the chamber	5 g cm^{-2}
Field size at the position of the reference point of the chamber	10 x 10 cm
Relative humidity	50%
Polarizing voltage and polarity	As in the calibration certificate
Dose rate	No reference values are recommended but the dose rate used should always be stated in the calibration certificate. It should also be stated whether a recombination correction has or has not been applied and if so, the value should be stated

In calibrating an ionization chamber or a dosimeter, as many influence quantities as manageable are kept under control. However, some influence quantities cannot be controlled, for example air pressure and humidity, or dose rate in ^{60}Co gamma radiation. The effect of these influence quantities can be taken into account by applying appropriate correction factors. Assuming that influence quantities act independently from each other, a product of correction factors is applied:

$$M_Q = M_Q^{uncorrected} \cdot \prod k_i \qquad (12)$$

where k_i refers to the different influence quantities.

6.2.1. *Air temperature and air pressure*

T_0 and P_0 are the reference conditions for chamber air temperature (in °C) and pressure. T and P are the actual air temperature (in °C) and pressure. Then in the user's beam, the correction factor for air temperature and air pressure $k_{T,P}$ is:

$$k_{T,P} = \frac{(273.2+T)}{(273.2+T_0)} \frac{P_0}{P} \qquad (13)$$

6.2.2. *Polarity effect*

Under identical irradiation conditions the use of potentials of opposite polarity in an ionization chamber may yield different readings. This phenomenon is called the polarity effect. If the used polarity differs from that at calibration, the following correction factor must be applied:

$$k_{pol}(V) = \frac{|M_+(V)| + |M_-(V)|}{2M} \qquad (14)$$

M_+ is the chamber signal obtained at positive chamber polarity.
M_- is the chamber signal obtained at negative chamber polarity.
M is the chamber signal obtained at the polarity used routinely (either positive or negative).

6.2.3. *Recombination effect*

Most important is an incomplete collection of charge in an ionization chamber cavity owing to the recombination of ions. Two separate effects take place: (i) the recombination of ions formed by separate ionizing particle tracks, termed general (or volume) recombination, which is dependent on the density of ionizing particles and therefore on the dose rate; (ii) the recombination of ions formed by a single ionizing particle track, referred to as initial recombination, which is independent of the dose rate. In pulsed radiation (i.e. at each linear accelerator!), the dose rate during a pulse is relatively high and general recombination is often significant.

In the IAEA Code of Practice it is recommended, that the correction factor k_s for pulsed beams be derived using the two voltage method:

$$k_s = a_o + a_1 \left(\frac{M_1}{M_2} \right) + a_2 \left(\frac{M_1}{M_2} \right)^2 \tag{15}$$

where the values of the collected charges M_1 and M_2 are measured at the polarizing voltages V_1 and V_2, respectively. V_1 is the normal operating voltage and V_2 a lower voltage. The ratio V_1/V_2 should ideally be equal to or larger than 3. The constants a_i are given in Table 8.

TABLE 8. Quadratic fit coefficients, for the calculation of k_s by the "two-voltage" technique in pulsed radiation, as a function of the voltage ratio V_1/V_2.

V_1/V_2	a_0	a_1	a_2
2.0	2.337	−3.636	2.299
2.5	1.474	−1.587	1.114
3.0	1.198	−0.875	0.677
3.5	1.080	−0.542	0.463
4.0	1.022	−0.363	0.341
5.0	0.975	−0.188	0.214

In continuous radiation, notably ^{60}Co gamma rays, the two voltage method may also be used and a correction factor derived using the relation:

$$k_s = \frac{\left(V_1/V_2 \right)^2 - 1}{\left(V_1/V_2 \right)^2 - \left(M_1/M_2 \right)} \tag{16}$$

6.3. APPROPRIATE VALUES FOR K_Q

Values for k_Q are dependent on the type of ionization chamber and on the quality of radiation (type, energy, machine). Appropriate values are provided in the IAEA Code of Practice TRS 398. In this Code practical recommendations and data for each radiation type have been placed in an individual section devoted to a particular radiation type:

Low energy X rays with generating potentials up to 100 kV and HVL of 3 mm Al (the lower limit is determined by the availability of standards)

Medium energy X rays with generating potentials above 80 kV and HVL of 2 mm Al

^{60}Co gamma radiation

High energy photons generated by electrons with energies in the interval 1–50 MeV

Electrons in the energy interval 3–50 MeV

Protons in the energy interval 50–250 MeV, with a practical range, R_p, between 0.25 and 25 g/cm^2

Heavy ions with Z between 2 (He) and 18 (Ar) having a practical range in water, R_p, of 2–30 g/cm^2

k_Q does not only depend on the type of radiation but additionally on its quality. The quality of radiation may be characterized by the fluence spectrum differential in energy. For practical reasons, a particular quantitative parameter, the so-called quality index defined for each category of radiation type is used. Accordingly, values of k_Q are tabulated as a function of this quality index. The selection of the correct value of k_Q therefore requires the determination of the quality index first. The methods to determine the quality index differs from one radiation type to another.

6.3.1. Definition of the quality index for HE photons

For high energy photons produced by clinical accelerators the beam quality Q is specified by the tissue phantom ratio $TPR_{20,10}$. This is the ratio of the absorbed doses at depths of 20 and 10 cm in a water phantom, measured with a constant SCD of 100 cm and a field size of 10 × 10 cm at the plane of the chamber. The most important characteristic of the beam quality index $TPR_{20,10}$ is its independence of the electron contamination in the incident beam.

6.3.2. Definition of the quality index for HE electrons

Quality indices used for megavoltage electron beam specification are commonly based upon:

The mean (average) electron energy of the incident spectrum striking the phantom surface

The half-value depth in water R_{50}

In the IAEA Dosimetry protocol the beam quality index for electron beams is the half-value depth in water R_{50}.

This is the depth in water (in $g\ cm^{-2}$) at which the absorbed dose is 50% of its value at the absorbed-dose maximum (Fig. 12). R_{50} is measured with a constant SSD of 100 cm, a field size at the phantom surface of at least 10 x 10 cm for $R_{50} \leq 7\ g\ cm^{-2}$ ($E_0 < 16$ MeV), and at least 20 x 20 cm for $R_{50} > 7\ g\ cm^{-2}$ ($E_0 \leq 16$ MeV).

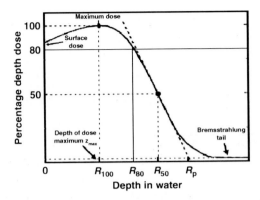

Figure 12. Illustration of several depth dose parameters for high energy electrons including the quality index parameter R50.

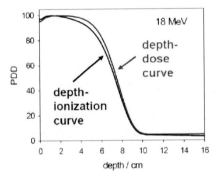

Figure 13. Difference of the depth dependence of ionization in air (the signal measured with an ionization chamber) and absorbed dose.

However, one must be cautious to derive the quality index R_{50} directly from the result obtained by an ionization chamber measurement. The reason is that relation between the absorbed dose in water and that in air is given by the mean mass stopping power ration water to air (see equation 7) which is not constant with the depth.

The difference between a normalized depth dose curve and the equivalent ionization depth curve in water is illustrated in Fig. 13.

Therefore, if measured depth ionization curves in water are to be directly used for the determination of R_{50}, it must be calculated using:

$$R_{50} = 1.029 \, R_{50,ion} - 0.06 \quad \mathrm{g\ cm}^{-2} \quad (R_{50,ion} \leq 10 \ \mathrm{g\ cm}^{-2})$$

$$R_{50} = 1.059 \, R_{50,ion} - 0.37 \quad \mathrm{g\ cm}^{-2} \quad (R_{50,ion} > 10 \ \mathrm{g\ cm}^{-2})$$

As an alternative to the use of an ionization chamber, other detectors (for example diode, diamond, etc.) may be used to determine R_{50}. In this case the user must verify that the detector is suitable for depth-dose measurements by test comparisons with an ionization chamber at a set of representative beam qualities.

6.4. CALIBRATION FACTORS FOR HIGH ENERGY ELECTRONS EXPRESSED AT DOSE MAXIMUM

Frequently one wants the basic output for an electron beam to be given at the depth of the dose maximum, z_{max}. This requires a conversion from the measured depth ionization curve to the depth dose curve and renormalizing to maximum dose. The conversion is performed by multiplying the depth ionization curve with the depth dependent water to air stopping power ratio adjusted to the beam quality of the electron beam using the following formula:

$$s_{w,a}^{\Delta}(z) = \frac{a + bx + cx^2 + dy}{1 + ex + fx^2 + gx^3 + hy} \qquad (17)$$

with $x = \ln(R_{50}/cm)$, and $y = z/R_{50}$

$a = 1.0752 \qquad b = -0.50867 \quad c = 0.08867 \; d = -0.08402$

$e = -0.42806 \quad f = 0.06463 \qquad g = 0.003085 \; h = -0.1246$

7. Cross calibration

7.1. CONCEPT

Cross-calibration refers to the calibration of a user chamber by direct comparison in a suitable user beam against a reference chamber that has previously been calibrated. A particular example is the cross-calibration of a plane-parallel chamber for use in electron beams against a reference cylindrical chamber calibrated in ^{60}Co gamma radiation. Despite the additional step, such a cross-calibration generally results in a determination of absorbed dose to water using the plane-parallel chamber that is more reliable than that achieved by the use of a plane-parallel chamber calibrated directly in ^{60}Co. The main reason is: problems associated with the perturbation correction for plane-parallel chambers in ^{60}Co, entering into the determination of k_Q, can be avoided.

7.2. DETERMINATION OF THE CROSS CALIBRATION FACTOR

The highest-energy electron beam available should be used; $E > 16$ MeV is generally recommended. Note: having performed the cross calibration, the electron beam used for that is now the new calibration quality!

The reference chamber (denoted by "ref") and the chamber to be calibrated (denoted by "X") are compared by alternately positioning each at the reference depth z_{ref} in water. When the charges of these to chambers (corrected for the different influence factors) are determined, the cross calibration factor

$$N^x_{D,w,Q_{cross}}$$

in terms of absorbed dose to water for the chamber under calibration at the cross calibration quality Q_{cross}, is given by:

$$N^x_{D,w,Q_{cross}} = \frac{M^{ref}_{Q_{cross}}}{M^x_{Q_{cross}}} N^{ref}_{D,w,Q_o} k^{ref}_{Q_{cross},Q_o} \tag{18}$$

7.3. SUBSEQUENT USE OF A CROSS-CALIBRATED CHAMBER

The cross-calibrated chamber with the new calibration factor may be used subsequently for the determination of absorbed dose in any arbitrarily other user beam of quality Q using the basic equation:

$$D_{w,Q} = M^x_Q \cdot N^x_{D,w,Q_{cross}} \cdot k^x_{Q,Q_{cross}} \tag{19}$$

where:

M^x_Q is the charge (corrected for all influence factors) measured with the cross-calibrated ionization chamber in an electron beam with quality Q

$N^x_{D,w,Q_{cross}}$ is the cross calibration factor obtained by equation 18

$k^x_{Q,Q_{cross}}$ is a new beam quality correction factor which now refers to the quality Q_{cross} (instead to that of ^{60}CO)

The values for $k^x_{Q,Q_{cross}}$ are derived using the procedure:

$$k^x_{Q,Q_{cross}} = \frac{k^x_{Q,Q_{int}}}{k^x_{Q_{cross},Q_{int}}} \tag{20}$$

where:

$$k^x_{Q,Q_{int}} \; ; \; k^x_{Q_{cross},Q_{int}}$$

are intermediate quality correction factors assuming that Q_{int} is used as the calibration quality. They are given in TRS 398, table 19. Although not really used in the measurement, this trick allows an easy calculation of the really required correction factor $k^x_{Q,Q_{cross}}$.

Appendix: Calculation formula for the quality correction factor k_Q

The values k_Q tabulated in TRS 398 have been obtained by calculation. A derivation for the calculation formula for k_Q can be based on the following:

Assume that one wants to determine absorbed dose in water directly by first principles, i.e. using a ionization chamber. Then the following steps referring to Fig. 14 must be performed:

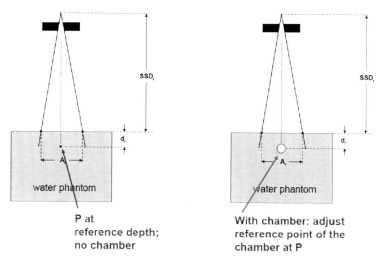

Figure 14. Illustration of the primary step in the determination of absorbed dose in water using an air cavity.

1. Step: Substitute the absorbing medium of water at the point of measurement by that of the air volume

   Step: Measure charge M and determine the dose in air using

   $$D_{air} = M/m_{air} \cdot \left(\overline{W}_{air}/e \right) \qquad (A1)$$

   Step: Translate the dose in air to that of water using:

   $$D_w = M/m_{air} \cdot \left(\overline{W}_{air}/e \right) \cdot s_{w,air} \cdot p \qquad (A2)$$

Then the problem arises: How one can get an accurate value for the mass m_{air} of the air of the sensitive volume of the ionization chamber?

Solution of the problem: Make use of the calibration of a chamber at a reference point in a reference quality Q_0 where the absorbed dose to water, D_{w,Q_0}, is exactly known.

In detail: Introduce a calibration factor N_{D,w,Q_0} such that:

$$N_{D,w,Q_0} = \frac{D_{w,Q_0}}{M_{ref}} \qquad (A3)$$

where M_{ref} is the charge measured at this calibration procedure in a calibration quality Q_0 under reference conditions. It follows that the absorbed dose under exactly these condition is given by:

$$D_{w,Q_0} = N_{D,w,Q_0} \cdot M_{ref} \qquad (A4)$$

At the same time we know that:

$$D_{w,Q_0} = \frac{M_{ref}}{m_{air}} \cdot \left(\overline{\frac{w}{e}}\right)^{Q_0} \cdot s_{w,air}^{Q_0} \cdot p_{Q_0} \qquad (A5)$$

where Q_0 always refers to the quality of the calibration beam. By equating these two expressions, M_{ref} cancels out and one obtains:

$$\frac{1}{m_{air}} \cdot \left(\overline{\frac{w}{e}}\right)^{Q_0} \cdot s_{w,air}^{Q_0} \cdot p_{Q_0} = N_{D,w,Q_0} \quad \text{or} \quad m_{air} = \left(\overline{\frac{w}{e}}\right)^{Q_0} \cdot s_{w,air}^{Q_0} \cdot p_{Q_0} \qquad (A6)$$

If this expression for the mass of air is filled in equation (A2), the absorbed dose can be determined by:

$$D_{w,Q} = M_Q \cdot N_{D,w,Q_0} \cdot \frac{\left(\overline{\frac{W}{e}}\right)^{Q} \cdot s_{w,air}^{Q} \cdot p_Q}{\left(\overline{\frac{W}{e}}\right)^{Q_0} \cdot s_{w,air}^{Q_0} \cdot p_{Q_0}} \qquad (A7)$$

Comparing this expression with equation (10) it follows that k_Q can be calculated by:

$$k_Q = \frac{\left(\overline{\frac{W}{e}}\right)^{Q} \cdot s_{w,air}^{Q} \cdot p_Q}{\left(\overline{\frac{W}{e}}\right)^{Q_0} \cdot s_{w,air}^{Q_0} \cdot p_{Q_0}} \qquad (A8)$$

For a detailed discussion on possible values for the parameters contained in equation (A8), the reader is referred to the appendix of the IAEA dosimetry protocol TRS 398.

References

1. ICRU International Commission on Radiation Units and Measurement. Fundamental Quantities and Units for Ionizing Radiation. ICRU Report 60, ICRU, Bethesda (1998).
2. IAEA International Atomic Energy Agency. "Absorbed Dose Determination in External Beam Radiotherapy: An International Code of Practice for Dosimetry based on Standards of Absorbed Dose to Water". Technical Report Series no. 398, IAEA, Vienna (2000).

NOMENCLATURE IN RADIOTHERAPY

GÜNTHER H. HARTMANN
German Cancer Research Center, Department of Medical
Physics in Radiation Oncology, Im Neuenheimer Feld 280,
D-69120 Heidelberg, Germany
g.hartmann@dkfz-heidelberg.de

Abstract In order to enable a meaningful comparison of therapeutic results within one hospital and/or between different institutions – concepts for a common language are essential. Such a concept is provided by the ICRU report 50. It recommends methods for specifying volumes and doses to be used for prescription, recording and reporting of a radiation treatment in external photon beam therapy. A supplement to the ICRU-50 Report is additionally treating the problem of internal and set-up margin encompassing the clinical target volume. Furthermore, the new useful parameter called the conformity index was introduced. Finally, the nomenclature for the specification of dose distribution is introduced using the concept of the ICRU reference point. The criteria how to select this point is demonstrated using simple examples of different treatment plans.

Keywords: Radiotherapy; aim of radiotherapy; volumes; recording; reporting; dose-volume-histogram

1.   Introduction

When performing a radiotherapy treatment, parameters such as the volume to which a specific absorbed dose was delivered, and the amount of absorbed dose as well as its distribution have to be specified in order to characterize the treatment. Characterization serves for different purposes:

- Prescription of therapeutic dose
- Recording of all parameters associated with the treatment
- Reporting

It is important that clear, well defined and unambiguous concepts and parameters are used in particular for reporting purposes to ensure a common language.

It is important that clear, well defined and unambiguous concepts and
parameters are used in particular for reporting purposes to ensure a common
language between different centers. This chapter is dealing with the main ideas
and concepts developed and published in the ICRU Reports 50 and 62 (Fig. 1).

The ICRU Reports 50 and 62 define and describe a terminology such as
target volume or volumes of critical structure that: (a) aid in the treatment planning process, (b) provide a basis for comparison of treatment outcomes. Most
important issues are:

- **Aim of therapy** which may have an impact on the choice of the volume to
  be treated
- **Volumes** which may be based on pure anatomical or geometrical concepts
  and which can be used to clearly indicate which volume is to be treated to
  the prescribed dose
- **Specification of dose** including standardized procedures to communicate
  the most significant dose or its most adequate heterogeneity
- **General Recommendations for Reporting Dose** under basic minimal
  requirements

Figure 1. The ICRU 50 and 62 reports.

2. Aim of therapy

It is important to define first the aim of therapy, since the determination of aim may have a direct impact on the choice of volume to be treated, the radiation dose, or the treatment technique. Three different aims of therapy can be distinguished:

- Radical treatment of a malignant disease: the aim of radical radiotherapy, also called "curative radiotherapy" is to decrease the number of tumor cells to a level that achieves permanent local tumor control. The volumes to be irradiated have to include any demonstrated tumor, and also volumes in which subclinical spread is expected but nor directly visible. Therefore, anatomical tumor borders may or may not be demonstrable. Furthermore, if the tumor has already been removed by surgery, the remaining tissue may contain subclinical disease; thus the limits of the target volume cannot be demonstrated clinically at all.
- Palliative treatment of malignant disease: the aim of palliative radiotherapy is to decrease various symptoms of a malignant disease in order to maintain an acceptable quality of the remaining life. This treatment may include all or only part of the demonstrated tumor. A frequent example is the irradiation of the spine for a painful deposit in a case of wide-spread metastases.
- Non-malignant Disease: this type of radiotherapy may or may not include all of the affected tissues. An example is the irradiation of dermatoses.

The relation between curative treatments and palliative treatments may be different in different countries. It is certainly true that in developing countries the number of palliative treatment is much more higher compared to curative treatments. In Germany the relation is almost evenly distributed with a number of 150,000 patients a year receiving a curative treatment and 100,000 patients a year receiving a palliative treatment.

3. Volumes

The process of determining the volume of a malignant disease to be treated to a prescribed dose consists of several distinct steps. These steps are normally included in the treatment planning procedure. During this procedure different volumes will be defined which are serving different purposes and therefore must take into account different attributes such as varying concentrations of malignant cells, probable changes in the spatial relationship between volume and beam during therapy, movement of patient, and possible inaccuracies in the treatment setup. The two most important different categories of volumes are (1) that of being based on pure anatomical or (2) geometrical concepts. Different coordinate systems are involved for that as shown in Fig. 2:

Two volumes should be defined prior to physical treatment planning: the Gross Tumor Volume (GTV) and the Clinical Target Volume (CTV). Later on other volumes such as Planning Target Volume (PTV), the Organs at Risk (OAR), and the Planning Organ at Risk Volume (PRV) have to be defined. As a result of treatment planning, the definition of further volumes such as the Treated Volume or the Irradiated Volume can be helpful. The following sections describe these different volumes that have been defined in the ICRU 50 report as principal volumes related to three-dimensional treatment planning. They are schematically illustrated in Fig. 3.

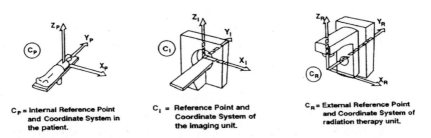

C_P = Internal Reference Point
and Coordinate System in
the patient.

C_I = Reference Point and
Coordinate System of
the imaging unit.

C_R = External Reference Point
and Coordinate System of
radiation therapy unit.

Figure 2. The definition of volumes in radiotherapy requires different types of coordinate systems. Whereas the patient and image coordinate system are used for anatomical volumes, a coordinate system of the radiation therapy unit is required for geometrical volumes.

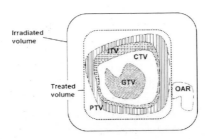

Irradiated volume

CTV

GTV

OAR

Treated volume

PTV

Figure 3. The set of volumes as defined in the ICRU Report 50 and their spatial relationship.

3.1. GROSS TUMOR VOLUME (GTV)

Definition: The Gross Tumor Volume (GTV) is the gross palpable or visible/demonstrable extent and location of malignant growth. The GTV is usually based on information obtained from a combination of imaging modalities (CT,

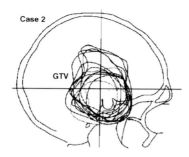

Figure 4. Variations in defining the GTV (brain tumor) based on a lateral radiograph. Eight radiation oncologists, two radio-diagnosticians, and two neurosurgeons have drawn the contours.

MRI, ultrasound, etc.), diagnostic modalities (pathology and histological reports, etc.) and clinical examination. There are several reasons to identify the GTV. (1) It is required for staging, (2) it provides a relation to the adequate dose to obtain local control, and (3) the regression of GTV may be used to predict tumor response.

The method used for the definition of the GTV and which is mostly image guided also influences the shape of the GTV. It must be therefore included in the description of the GTV. However, even using the same technique, inter-observer variations may be significant. An example is shown in Fig. 4.

3.2. CLINICAL TARGET VOLUME (CTV)

Definition: the Clinical Target Volume (CTV) is the tissue volume that contains a demonstrable GTV and/or sub-clinical microscopic malignant disease, which has to be eliminated. This volume thus has to be treated adequately in order to achieve the aim of radical therapy. The CTV normally includes the GTV. Further aspects are: the CTV often includes the area directly surrounding the GTV that may contain microscopic disease and also other areas considered to be at risk and therefore require treatment such as positive lymph nodes.

The CTV is an anatomical-clinical volume. It is usually determined by the radiation oncologist, often after other relevant specialists such as pathologists or radiologists have been consulted. The CTV is usually stated as a fixed or variable margin around the GTV. In some cases the CTV is the same as the GTV. An example for that is the prostate irradiation which is directed to the gland only. Also several non-contiguous CTVs can be defined that may require different total doses to achieve treatment goals. An example is the boost therapy.

In Fig. 5 the use of two CTVs is illustrated for the case of a lung tumor.

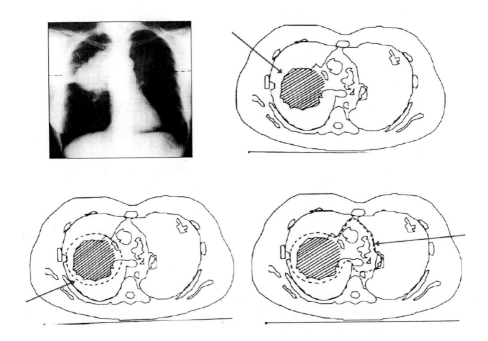

Figure 5. Example of a lung tumor where two CTVs were joined together. In addition to the CTV 1 surrounding the tumor (representing the GTV, upper right) no medestinal lymph nodes metastases could be demonstrated by clinical investigations. However, the lymph nodes as well as the parts of the contra-lateral hilar region are indeed considered to be at high risk thus forming a CTV 2. Both CTVs will be treated with the same dose and beam. Consequently, they are joined together shown as dotted line, lower left. (Pictures taken from ICRU Report 50.)

3.3. PLANNING TARGET VOLUME (PTV)

Once the CTV has been defined, and a specific external beam radiation treatment has been selected as the treatment modality, a suitable arrangement of beams must be selected in order to achieve the adequate dose distribution. In contrast to the CTV, which is an anatomical concept, the Planning Target Volume (PTV) is a geometrical concept.

It is defined to select appropriate beam arrangements, taking into consideration the net effect of all possible geometrical variations, in order to ensure that the prescribed dose is actually absorbed in the CTV. The PTV requires an appropriate margin to the CTV Margins can be non-uniform but should be three dimensional. For moving CTV a reasonable way of thinking would be: "Choose margins so that the target is in the treated field at least 95% of the time."

The example of lung tumor used in Fig. 5 is again used to demonstrate the relation between CTV and PTV in Fig. 6.

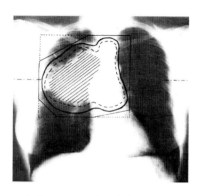
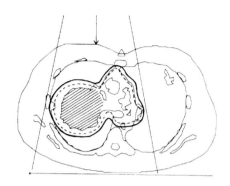

Figure 6. Same example of a lung tumor as used in Fig. 5. In relation to the coordinate system of the radiotherapy unit (indicated by the field borders, right) the CTV will move with respiration, and the patient as a whole will not perfectly immobilized during each fraction. Further setup uncertainties will occur. Therefore, the treatment has to be planned for a larger volume than the CTV. A suitable PTV (indicated by the thick solid line) is defined in the X-ray image left as well as in the axial CT section right. (Picture taken from ICRU Report 50.)

3.4. MARGINS

The difference between the shape of a CTV and that of a PTV is normally characterized by a margin. In order to determine the required margins, it is useful to define margins separately dependent on the source of uncertainty of the shape and position of a CTV, namely organ motion and patient setup. According to ICRU 62, organ motions are conceptually included in so-called internal margins, whereas uncertainties in the patient set-up (random and systematic) are included by external margins.

3.4.1. *Internal Margin and Internal Target Volume*

A particular margin (IM) must be added to CTV to compensate for physiologic movements and variations in size, shape and position of the CTV during therapy in relation to an internal reference point. Internal reference points are usually defined by the bony anatomy. Variations in size, shape and position of the CTV may be due to organ motions such as breathing, bladder or rectal contents, etc. This type of margin is referred to as the Internal Margin. CTV and IM together form the Internal Target Volume (ITV).

3.4.2. *Set-up Margin (SM)*

To account for uncertainties in patient positioning and alignment of the beams during treatment planning and through all treatment sessions, a further Set-up Margin (SM) for each beam is needed. The SM is referenced in the external coordinate system.

The PTV thus includes the internal target margin and the additional margin for set-up uncertainties, machine tolerances, and intra-treatment variations.

3.5. TREATED VOLUME

Due to limitations of irradiation techniques, the volume receiving the prescribed dose does not generally match the PTV. This leads to the concept of Treated Volume. The Treated Volume (TV) is the tissue volume that is planned to receive at least a dose selected and specified as being appropriate to achieve the purpose of treatment. Therefore, the TV is the volume enclosed by an isodose surface.

A conformity index CI can be employed when the PTV is fully enclosed by the TV. In this case the quotient of TV and PTV can be determined and used as the CI.

$$CI = \frac{\text{Treated Volume}}{\text{PTV}} \qquad (1)$$

The conformity index is normally greater than 1.

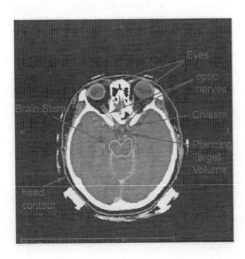

Figure 7. Examples of different organs at risk in the vicinity of a target volume in the brain such as brain stem, eyes, optic nerves, or chiasm.

3.6. ORGAN AT RISK

Organs at risk (OAR) are normal tissues whose radiation sensitivity may significantly influence treatment planning and/or prescribed dose (Fig. 7). Presently, our knowledge about the sensitivity of normal tissue is derived mainly from clinical observations.

In the same way as with the Planning Target Volume, any movement of the Organ at Risk during treatment, as well as uncertainties in the set-up during the whole treatment course, must be considered. This leads, in analogy with the PTV, to the concept of Planning Organ at Risk (PRV).

4. Recommendations for recording and reporting

The aim of the recommendation of ICRU Report 50 is to promote uniformity between different radiotherapy centers, when reporting their treatments. For this purpose it is essential that the same type of treatment be reported in the same way, using the same terminology and definitions.

When reporting a radiation treatment, the following oncological information must be given first:

- Description of the Gross Tumor Volume
- Description of the Clinical Tumor Volume(s)

This step corresponds with the primary aim of the clinical work-up of a patient with a malignant disease which is to define the site of the primary tumor, its size and possible invasion of adjacent structures, and to detect regional lymph node involvement and distant metastases. This staging procedure should result in an accurate assessment of the extent of the disease. This can be accomplished (and it is also recommended) using plain language. However, several systems for coding anatomy and staging are also available.

Next to the recording and reporting of the oncological information, the information concerning the treatment itself must be given. It consists of:

- Planning Target Volume
- Treated Volume
- Irradiated Volume
- Planning Organ at Risk Volume

The recommendations concerning the PTV are that the PTV is described by giving the size of the margins around the CTV in all relevant directions. Table 1 gives an example for recording and reporting a treatment of a lung cancer (excluding dose prescription and reporting).

TABLE 1. Example of using the recommendations on recording on reporting with the treatment of a lung cancer (excluding dose prescription and reporting).

Item	Description
Clinical situation	65-year-old female smoker, presented with persistent cough. Chest radiography showed a right hilar mass. Bronchoscopy showed endobronchial tumor in the right main bronchus. Biopsy revealed a squamous cell carcinoma. Ct confirmed the lesion, plus a lymphad-enopathy at the right hilum. No evidence of mediasinal lymphadenopathy. Clinical stage IIB (T2 N1 M0)
Aim of therapy	Radical radiotherapy
GTV	1. Primary endobronchial tumor 2. Hilar lymphadenopathy
CTV	The combined CTV includes GTV 1 + GTV 2 + local subclinical extensions
PTV	A 10 mm margin was added to the CTV
Organs at risk	A: spinal cord; B: left lung; C: heart; D: oesophagus
Beam arrangement	Four isocentric beams
Technique	18 MV photon beams
Patient positioning	Supine with head on standard head rest and arms above head. Alpha cradle immobilization. Laser alignment and skin, cradle marks

5. Dose specification and prescription

The complete prescription of radiation treatment must include a definition of the aim of therapy, the volumes to be considered, and a prescription of dose and fractionation. Only detailed information regarding total dose, fractional dose and total elapsed treatment days allows for proper comparison of outcome results. Different concepts have been developed to characterize the dose and its distribution within a volume.

5.1. THE ICRU REFERENCE POINT

It is in particular important to document the dose in the planning target volume. For this purpose a point at or near the center of the PTV appears to be best suited. The ICRU 50 recommendation is based on the selection of a certain point within the CTV, which is referred to as the ICRU reference point. It is located at a point chosen to represent the delivered dose using the following criteria:

- The point should be located in a region where the dose can be calculated accurately (i.e., no build-up or steep gradients).
- The point should be in the central part of the PTV.
- For multiple fields, the isocenter (or beam intersection point) is recommended as the ICRU reference point (Fig. 8).

Specific recommendations are made with regard to the position of the ICRU (reference) point for particular beam combinations:

- For a single beam: the ICRU point is located on central axis at the center of the target volume.
- For parallel-opposed equally weighted beams: the ICRU point is located on the central axis midway between the beam entrance points.
- For parallel-opposed unequally weighted beams: the ICRU point is located on the central axis at the centre of the target volume.
- For other combinations of intersecting beams: the ICRU point is located at the intersection of the central axes (insofar as there is no dose gradient at this point).

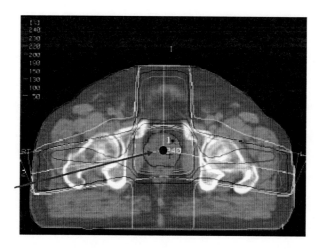

Figure 8. ICRU reference point for multiple fields. Example for a three field prostate boost treatment with an isocentric technique. The ICRU (reference) point is located at the isocenter.

5.2. SIMPLE PARAMETERS FOR THE SPATIAL DOSE DISTRIBUTION

When a dose to a given volume has been prescribed, then the corresponding delivered dose should be as homogeneous as possible. However, some heterogeneity has always to be accepted due to technical and clinical reasons. Thus, when prescribing the dose, one has to foresee a certain degree of heterogeneity.

It is recommended that the best technical and clinical conditions should be kept within +7% and −5% of prescribed dose.

Simple parameters to characterize the spatial dose distribution are:

- The maximum dose within the PTV as well as at tissues outside the PTV. In almost all cases the volume to be taken into account for that should not be smaller that 15 mm in diameter.
- The minimum target dose.
- The average dose.
- The median dose.
- Hot spots; a hot spot represents a volume outside the PTV which receives a dose larger than 100% of a specified PTV dose. Again, the volume to be taken into account for that should not be smaller that 15 mm in diameter.

5.3. DOSE VOLUME HISTOGRAMS

Dose volume histograms (DVHs) summarize the information contained in a three-dimensional treatment plan. This information consists of dose distribution data over a three-dimensional matrix of points over the patient's anatomy. DVHs are extremely powerful tools for quantitative evaluation of treatment plans. In its simplest form a DVH represents a frequency distribution of dose values within defined volumes such as the PTV itself or a specific organ in the vicinity of the PTV.

Rather than displaying the frequency, DVHs are usually displayed in the form of "percent volume of total volume" on the ordinate against the dose on the abscissa. Two types of DVHs are in use: (1) direct (or differential) DVH, and (2) cumulative (or integral) DVH using the procedure to plot the volume which receives at least the given dose versus dose.

5.3.1. *Direct (Differential) Dose Volume Histogram*

To create a direct DVH, the computer sums the number of voxels which have a specified dose range and plots the resulting volume (or the percentage of the total organ volume) as a function of dose.

The ideal DVH for a target volume would be a single column indicating that 100% of the volume receives the prescribed dose. For a critical structure, the DVH may contain several peaks indicating that different parts of the organ receive different doses. An example illustrating differential DVHs for prostate cancer and rectum is shown in Fig. 9.

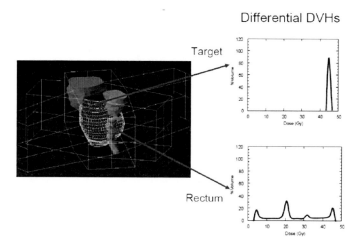

Figure 9. Illustration of differential Dose Volume Histograms for a prostate cancer and the rectum.

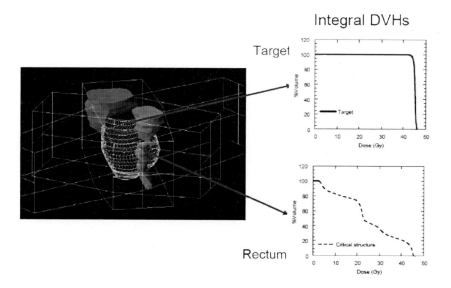

Figure 10. Illustration of cumulative Dose Volume Histograms for a prostate cancer and the rectum.

5.3.2. *Cumulative (Integral) Dose Volume Histogram*

Traditionally, physicians have sought to answer questions such as: "How much of the target is covered by the 95% isodose line?" In 3-D treatment planning this question is equally relevant and the answer cannot be extracted directly from the direct (differential) DVH, since it would be necessary to determine the area under the curve for all dose levels above 95% of the prescription dose. A better answer can be provided by an cumulative DVH. To create a cumulative DVH, the computer calculates the volume of the target (or critical structure) that receives at least the given dose and plots this volume (or percentage volume) versus dose (Fig. 10). All cumulative DVH plots start at 100% of the volume for zero dose, since all of the volume receives at least no dose.

While displaying the percent volume versus dose is more popular, it is also useful in some circumstances to plot the absolute volume versus dose. For example, if a CT scan does not cover the entire volume of an organ, such as the lung and the un-scanned volume receives very little dose, then a DVH showing percentage volume versus dose for that organ will be biased, indicating that a larger percentage of the volume receives dose.

Although very popular for modern three-dimensional treatment planning, there is, however, a drawback of the DVHs which is the loss of spatial information that results from the condensation of data when DVHs are calculated.

References

1. ICRU International Commission on Radiation. Prescribing, recording, and reporting photon beam therapy. ICRU Report 50, ICRU, Bethesda (1993).
2. ICRU International Commission on Radiation. Prescribing, recording, and reporting photon beam therapy (supplement to ICRU Report 50). ICRU Report 62, ICRU, Bethesda (1999).

QUALITY ASSURANCE IN RADIOTHERAPY

ALAN McKENZIE
Bristol Oncology Centre, Hortfield Road, Bristol, UK
alan.mckenzie@ubht.swest.nhs.uk

Abstract A common feature of the Radiotherapy Centres where there have been major accidents involving incorrect radiotherapy treatment is that they did not operate good Quality Assurance systems. A Quality Assurance system is sometimes called a Quality Management system, and it is designed to give assurance that quality standards are being met. One of the "spin offs" from operating a Quality Management system is that it reduces the likelihood of a radiotherapy accident. A detailed account of how to set up a quality system in radiotherapy has been given in an ESTRO booklet.[2]

Keywords: Quality Assurance systems; Quality Control; protocols; target delineation; introduction of a Quality System; organisation of a Quality System; control of a Quality System

1. Introduction

Among the quality standards that must be met in a Radiotherapy Centre are dosimetry standards and geometric standards. Meeting these standards is a necessary (but not complete) condition for good radiotherapy. It is widely accepted that dose discrepancies in the order of 6% or less will not be clinically detectable in individual patients. If this limit is not to be exceeded in more than one patient in 20, this represents two standard deviations, which means that the dosimetry standard should aim to deliver dose with an uncertainty not exceeding 3%. There is a chain of calibrations and calculations that lead to the final delivery of dose in a patient. The uncertainties in the component links in that chain (primary standard calibration, field chamber intercomparison, calculation of dose in a patient etc.) are each of the order 1–2%, so that the quadrature sum representing the overall uncertainty is close to the required 3%, or at least does not significantly exceed it. The Quality System is designed to ensure that the activities of each link in the chain are performed to the required tolerance.

Sometimes standards are required to ensure that target volumes are covered by radiation beams as accurately as practicable, with minimal risk to normal organs and tissues. Margins are drawn around target volumes which are supposed to be large enough to account for the principal sources of uncertainty: target delineation, set-up uncertainty, physical accuracy of radiotherapy equipment including "machine geometry" and motion of the target volume within the patient. Uncertainties in target delineation is generally the largest of these uncertainties,[1] but the aim should be to minimise the set-up and machine geometry uncertainties over which there is potentially more control.

Most machine parameters, such as light and X-ray field size, laser alignment etc. drift slowly over time and, in practice, the discrepancies are not adjusted back to zero until an action level is reached, which is typically 2 mm. The likelihood of finding any given discrepancy is therefore approximately constant for all deviations up to the action level so that the probability density function may be represented by a "top hat" distribution. The standard deviation of such a distribution (with an action level of 2 mm) is $2/\sqrt{3}$ mm[1] which is approximately 1 mm. This is acceptably small compared with the standard deviations of set-up, delineation and organ motion and is therefore a suitable geometric standard for this Quality System.

It is important to differentiate between the terms "Quality Assurance" and "Quality Control". Quality Control refers to the actions required to ensure that particular standards are being met, and this includes measuring output to check that it is within 2% of the standard, or checking that the lasers are accurate to within 2 mm. Quality Assurance refers to use of the actions that are necessary to give assurance that such quality control procedures take place. So Quality Assurance is a management system and Quality Control is a sub-set of the system. Other sub-sets include, for example:

- Writing protocols for the QC checks
- Managing these protocols to ensure that everyone has an up-to-date copy
- Ensuring that personnel are trained to carry out the QC procedures
- Analysing the training needs of the personnel and keeping training records
- Arranging systems to identify out-of-date conditions
- Drawing up timetables of machine QC
- Managing absence so that assurance can be given that there will always be personnel available to carry out the checks
- Drawing up a programme (organisations charts of personnel) so that responsibilities of individuals are clearly defined
- Making provision for resources to fund the equipment and personnel required to meet the quality standards etc.

2. Advantages of a Quality System

It is important that the advantages of a Quality System are clearly understood. The advantages include:

1. A Quality System is a management system that can cover every aspect of the Radiotherapy Centre or the department to which it applies.

2. A Quality System ensures continuing quality improvement by requiring management to seek out best practice from peer centres or from the national and international literature.

3. A Quality System brings about a culture change so that all personnel in the department feel they own the system and are pro-active in anticipating changes of practice by applying the principles of the Quality System in advance of the new practice being introduced. This way, personnel are not taken by surprise by the introduction, say, of gated radiotherapy, but have already worked out what will be necessary to install, commission and operate the system in a safe way in clinical practice.

4. A Quality System raises the morale of staff because it requires communication between different groups in order to ensure best and safest practice in all radiotherapy activities. Such communication often acts to defuse misunderstandings between groups and leads to a sense of "corporate identity".

5. A Quality System ensures efficiency by removing out-of-date protocols and obsolete practices and clearly states where information, documents and data are to be found, resulting in faster and more accurate responses to situations as they arise.

6. It follows from the above that a Quality System must reduce the likelihood of errors, although this is difficult to prove, as it would require comparing two Centres, one with a Quality System and one without. The Centre without the Quality System may have a higher error rate, but these errors, if they are not substantial, may well go undetected, because they do not have the Quality System in place that helps to detect errors.

7. A Quality System reduces the chance of litigation, partly because of the reduced likelihood of errors and partly because lawyers will perceive that the chances of winning a claim are reduced when a Centre can demonstrate a well functioning Quality System.

It has been observed that in all areas of activity, not restricted to medicine, when one organisation adopts a Quality System formally, then neighbouring or rival organisations will do the same. The presence of an externally accredited Quality System inspires confidence and is seen as a way to attract patients who might otherwise choose to go elsewhere for their treatment.

3. Introduction of a Quality System

There are four phases in introducing a Quality System into a Centre:

- Preparation
- Development
- Implementation
- Consolidation

3.1. PREPARATION

First, a small team must be chosen who will be looking to the future of the Quality System. This team will include a Quality Manager who should have sufficient seniority, authority and support from the head of the organisation to be able to implement the necessary strategies.

The department must be informed that the organisation is embarking on Quality Assurance. Some personnel will be unhappy with this news and may even actively resist the implementation, because they perceive the Quality System as removing some of the mystique surrounding their job by reducing it to a list of protocols. This perception is wrong and should be anticipated by bringing in people who are experienced in Quality Systems in other Centres to make presentations and answer questions during the informing phase. Part of the preparation will be to design the structure of the Quality System which will usually have a fairly brief statement of the quality policy (see below), a larger section containing procedures and protocols, and an extensive amount of detailed work instructions, forms and data, all in a format that enables the Quality System to locate, update and control them. As part of the preparation, an inventory of the existing structure is made, which will include existing data and work instructions (but which may require to be given new identification tabs etc.).

3.2. DEVELOPMENT

At this stage, the policy (see below) of the organisation is defined and agreed, and procedures and work instructions are written. The best people to do this, of course, are the same personnel who carry out these activities, and this also gives them "ownership" of part of the Quality System, so that they become pro-active in anticipating changes and accommodating such changes with the Quality System in the future. It is best to adopt a minimalist approach at this stage – only essential procedures and work instructions should be produced. It is a common mistake on the part of institutions adopting a Quality System for the first time, to write too much, which then becomes an administratively larger task to control.

3.3. IMPLEMENTATION

Once the documents of the Quality System have been put in place, personnel are then trained to use it. As the system is implemented, it will become clear whether any parts of the system need to be improved or, indeed, whether anything has been forgotten. This is termed "validation".

3.4. CONSOLIDATION

Once the system has been running for, say, 6 months, an audit can be made of the system. This involves reading the work instructions and procedures and then checking to see if all of the documented activities are, indeed, being carried out. The audit should be done "internally" at this stage, that is, using personnel who belong to the organisation and who are themselves covered by the Quality System.

The results of this audit are then analysed by a management team who review the system. These "system reviews" and "management reviews" will be conducted henceforth on a regular basis, say, every 6 months and, apart from the results of internal audit, all other aspects of the Quality System are checked and improvements made as necessary (see below).

4. Organisation of a Quality System

The contents of a Quality System Manual will typically include the following:
1. Introduction
2. Aims and policy of the organization
3. Structure of the organization
4. Means and materials
5. Process control
6. Knowledge and skills
7. Control of the Quality System

4.1. AIMS AND POLICY OF THE ORGANISATION

Typically, these aims and policy might be:
- To provide sufficient cancer care to the population which the Centre has to serve
- To be the Centre for specialised techniques
- To be the training hospital for the region
- To be networked with other Cancer Centres
- To be a high-quality Cancer Centre

As part of this process, the Quality System should document:

- Who is responsible for keeping the aims and policy up-to-date
- How data (e.g., patient statistics) are collected, analysed and presented
- Who commissions (i.e., formally requires) the provision of cancer care (government, private suppliers) and who will negotiate with the commissioners

4.2. STRUCTURE OF THE ORGANISATION

An organogram (organisational chart) should be drawn up so that the responsibilities of all personnel are clear. This must reflect the agreed job descriptions for individual personnel. The Quality System must state how those responsibilities are reviewed and how the performance of personnel is assessed. A list of essential committees should be prepared, with the membership and terms of reference indicated. There must be an infrastructure for communicating between different sections in the organisation and this should be clarified, possibly through the committee structure.

In order to avoid unnecessary meetings, and wasting people's time, each meeting should contain as its terms of reference the following:

1. The aim of the meeting
2. The membership
3. Who represents individual Sections
4. What decisions can be made
5. Frequency of the meeting
6. How decisions are recorded
7. How this information is transmitted to those not present

4.3. MEANS AND MATERIALS

The Quality System should contain a list of the equipment infrastructure, such as linear accelerators, CT scanners, simulators etc., as well as a description of responsibilities attached to this equipment. These responsibilities will include the process for planning future needs of the department, including an equipment replacement programme and how to proceed with procuring new equipment. They will also include procedures for:

1. Accepting, commissioning and maintaining equipment
2. Extending equipment life and replacement as necessary
3. Action in case of unexpected breakdown
4. Equipment quality control

5. Control of a Quality System

5.1. PROCESS CONTROL

At the top level, the Quality System should describe the "patient pathway", including all of the processes that affect the patient from the first time the department is consulted about a patient to the moment when a patient is discharged from follow up (which may be several years).

For each process the procedure should include the following headings:

1. Objective (e.g., to measure linear accelerator output)
2. Scope (e.g., 6 mV photons only)
3. Responsibilities (e.g., Head of Radiation Dosimetry)
4. Method (e.g., IAEA protocol)
5. Documents (relevant publications and references to other parts of the Quality System)

5.2. KNOWLEDGE AND SKILLS

A statement must be made on the knowledge, qualifications and experience required for all personnel and arrangements for their continuing professional education. It is important to document the requirement to train personnel when the techniques and technologies are to be introduced.

Job descriptions must accurately reflect these requirements. It is useful to have a policy on how experience gained at meetings and conferences outside the department is cascaded to relevant personnel.

5.3. CONTROL OF THE QUALITY SYSTEM

The Quality System is reviewed by managers regularly.

The following diagram illustrates the process (Fig. 1).

Concessions are made when it is deemed necessary not to follow the detailed procedures of the Quality System for an exceptional circumstance. For instance, a concession form may be completed to allow use of an ionisation chamber calibration factor which is out-of-date because the primary standard laboratory has had a problem in providing calibrations. The person signing the concession form would typically be the Head of Radiotherapy Physics in this instance. He/she would take into account results of the recent Strontium-90 checks on the ionisation chamber, and whether other ionisation chambers indicated similar outputs on linear accelerators that had been recently calibrated with the ionisation chamber.

THE MANAGEMENT REVIEW CYCLE

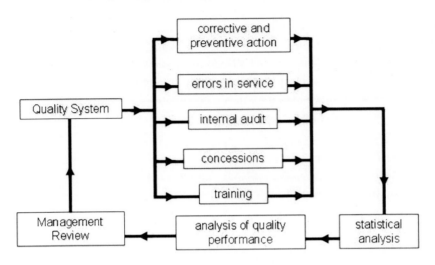

Figure 1. The Quality System reviewing.

Processes that must be in place to control the Quality System include:

1. Document and data control
2. Procedure writing (i.e., writing procedures formally using the headings listed earlier)
3. Quality records
4. Management review of quality standards
5. Analysis of quality performance
6. Corrective and preventive action
7. Incidents
8. Concerns
9. Complaints
10. Internal quality audit

A good Quality System should have no difficulty in successfully passing an external quality audit by an approved organisation. Such external audits must be paid for but the motivation that such audits inspire to keep the Quality System in good working order is well worth the expense. It is also a demonstration to the outside world that the organisation has achieved the appropriate standard.

References

1. British Institute of Radiology 2003. Geometric Uncertainties in Radiotherapy: Defining the Planning Target Volume, British Institute of Radiology, London, UK: British Institute of Radiology, 2003.
2. Leer JWH, McKenzie AL, Scalliet P and Thwaites DI 1998. Practical Guidelines for the Implementation of a Quality System in Radiotherapy. ESTRO booklet no. 4, Brussels.

RADIOTHERAPY ACCIDENTS

ALAN McKENZIE
Bristol Oncology Centre, Hortfield Road, Bristol, UK
alan.mckenzie@ubht.swest.nhs.uk

Abstract A major benefit of a Quality Assurance system in a radiotherapy centre is that it reduces the likelihood of an accident. For over 20 years I have been the interface in the UK between the Institute of Physics and Engineering in Medicine and the media – newspapers, radio and TV – and so I have learned about radiotherapy accidents from personal experience. In some cases, these accidents did not become public and so the hospital cannot be identified. Nevertheless, lessons are still being learned.

Keywords: Quality Assurance system; likelihood of accident; risk analysis; ROSIS

1. Radiotherapy accidents

My first case was one of the largest radiotherapy accidents anywhere in the world. It involved 207 patients who were all given a dose that was 25% greater than intended. It happened in 1988 when a new cobalt-60 source was installed into a treatment unit in the radiotherapy centre in Exeter, UK. The physicist measured the output for a shorter period than normal (0.4 min instead of 0.5 min). However, in converting to dose rate, instead of dividing the measured dose by 0.4, he divided by 0.5, because he was used to doing that. Therefore, he underestimated the dose rate, and so patients were given doses which were 25% higher than prescribed.

The error was made worse because early reports of excessive skin reactions on the patients were not fully explained. A good Quality Assurance system would have specified:

1. In major calibrations (such as a source change or after a new monitor chamber is fitted on a linear accelerator), two physicists should determine the output independently and both should sign to show that they have done this.
2. The radiotherapy centre should take part in an external audit that includes independent assessment of output.

3. Treatment should not proceed until all apparent discrepancies are fully explained (in this case, the unexplained observation was the excessive skin reaction to the treatment).

An accident involving an even greater number of patients (1,045 in total) was discovered in 1990, although it had been in progress steadily for the previous 10 years. In 1980, in the radiotherapy centre in Stoke-on-Trent, near the centre of England, the physicist installed a new treatment planning system. Up until that time, no patients had been treated isocentrically: all patients had fixed-SSD beams. It was decided to use the new treatment planning system to treat some patients isocentrically, which meant using beams at less than 100 cm SSD. Since the radiation beams had been calibrated at 100 cm SSD, the physicist agreed with the radiographers (radiation technologists) that the calculated monitor units should be reduced by the inverse-square ratio on isocentric beams. Unfortunately, the new treatment planning computer already contained an algorithm to account for this inverse-square ratio, and so the consequence was that patients received a dose which was too low when they were treated isocentrically.

Over the 10 years from the time when the treatment planning system was installed, only 6% of patients were treated isocentrically and so there was no clear pattern of loss of tumour control in these patients. Indeed, the mistake was discovered by the same physicist who made the error when the treatment planning system was replaced and she noticed the discrepancy when commissioning the new system.

A Quality System would have reduced the likelihood of this error happening, because it would have specified:

1. The commissioning of new equipment such as treatment planning systems and radiotherapy treatment machines should be checked independently by a second physicist.
2. Major changes in technique should be verified, where possible, by measurements in a phantom.
3. There should be participation in external audit, which should include measuring a dose given according to calculation by the centre's treatment planning system.
4. Staff involved with new equipment should receive appropriate training and should fully understand the use of the new equipment before it is used in clinical practice.

A third accident that involved many patients in a UK radiotherapy centre occurred in 2002 in Plymouth. A total of 132 radiotherapy breast patients were given the wrong dose of radiation because the physicist made a mistake when devising an algorithm to improve the monitor unit calculation by the treatment planning system. The intention was to remove the small errors which occur when treatment planning systems are used to calculate dose to reference points close to the skin surface in grazing-incidence beams treating the breast or chest wall. Unfortunately, the algorithm relied on a mistaken understanding of the system for expressing monitor units in a Varian dynamic wedge and, as a result, patients received less radiation then intended. A Quality System would have prevented the accident in the same way that the Stoke-on-Trent accident would have been prevented.

In an accident involving only one patient, but which was headline news in the UK, a patient in Glasgow received an overdose of nearly 70% to the brain. This happened because the centre had two systems for transmitting monitor units to the linear accelerators. In the normal system, the number of monitor units for the full dose per fraction would be transmitted but, for CNS treatments, the rule was that the monitor units for a normalised dose of 1 Gy would be transmitted, and the radiotherapy technologists would multiply the monitor units by the number of grays which had been prescribed. Unfortunately, the physicist forgot to reduce the monitor units to the number for 1 Gy, but the radiotherapy technologist continued to multiply the monitor units by the prescribed dose in grays according to protocol. So, when a dose of 1.67 Gy was prescribed, the patient received a dose that was 67% too high.

A Quality System requires a risk analysis of any new technique before it is used in clinical practice. So, when the centre in Glasgow commissioned the system of calculating the monitor units for the prescribed dose rather than for the normalised dose of 1 Gy, it should have been recognised that considering CNS to be a special case was a high-risk strategy. A Quality System would also have ensured that staff were properly trained to cope with the double system and that they understood their responsibilities in planning and checking. The Quality System would also require regular inspection and appropriate revision of all work instructions (so that the CNS anomaly might not have been overlooked).

In many accidents, a good system of in-vivo dosimetry for most patients at the beginning of their treatment would detect the problem and prevent the error propagating throughout the complete course of treatment.

In one further case in the UK, the identity of the centre must be kept confidential, but there is still a lesson to be learned. The accident happened in a centre where treatment details were entered, at that time, into the linac network by hand, using hardcopy from the treatment planning system. In the case in question, an operator forgot to specify that a wedge should be in place in a glancing-beam treatment of a chest wall. As a result, a high number of monitor units were delivered to the chest wall unattenuated by the metal wedge and so the patient received 94 Gy in 14 fractions instead of the intended 40 Gy in 15 fractions. On each of the 14 occasions when the patient was treated, radiotherapy technologists failed to notice that the system driving the linear accelerator was declaring that the wedge was not in position and they proceeded with treatment each time. This automatic assumption that the system is normal has been called "involuntary automaticity"[1] and it should not be underestimated as a potential risk in any system that requires a response from an operator to signal that treatment may proceed. Much can be learned about the causes of accidents by studying individual accounts. An excellent collection of the essential details of accidents up to the year 2000 is to be found in the IAEA publication "Lessons learned from accidental exposures in radiotherapy".[3] This IAEA publication lists the following causes for accidents arising from mistakes in creating and developing the scientific framework of radiotherapy, which therefore result in multiple-patient accidents (see Table 1).

TABLE 1. Generic cause of multiple-patient accidents.

Generic cause	No of accidents
Calibration and use of radiation measurement system	4
Commissioning/calibration of treatment Unit	15
Commissioning and use of treatment planning system	11

Unfortunately, since this work was published, there have been further major accidents in all three categories.

It must be understood that it is unlikely that a major accident will occur in any given radiotherapy centre. There will be many small errors and "near misses" and the centre should collect these data and analyse them for trends. However, the larger the number of such incidents, the more likely it is that there could be a much more serious error in such a centre.

Typical causes of small errors and near misses to be expected in a centre are listed below:

- Calculation of monitor units
- Treatment/distance
- Wedge (presence/absence/orientation)
- Field centre (asymmetry/orientation)
- Shielding
- Patient positioning
- Patient identification
- Prescription miscommunication

It is possible to evaluate the case for quality control checks in terms of the probability that patients will be harmed or lives saved. On the one hand, the total cost, in terms of costs of personnel, will increase as the number of checks increases. On the other hand, the benefit in terms of human life saved will increase. Such a process is called optimization, and the UK Institute of Physics and Engineering in Medicine (IPEM) has produced a report on the subject.[2] The broad conclusion of this report is that the frequencies of checking which have been established empirically (for instance, checking linear accelerator output once per week, with a daily constancy check) are already close to optimal.

Finally, the importance of selecting staff with the necessary aptitude for physics cannot be over-emphasised, and neither can the necessity of ensuring that these staff continue to receive appropriate training throughout their working lives. There is a warning in the IAEA report that says that "a shortage of staff usually leads to an increase in errors. In addition, a reassessment of staff and training has to be undertaken when new equipment is purchased or when a new technology or treatment modality is introduced."

While this lecture focuses mainly on the circumstances that underlie the major errors – those that reach the headlines in newspapers and television – it should be appreciated that the number and type of "near misses" that happen in a radiotherapy centre are indicative of the likelihood of a major error occurring. Therefore, departments should collect reports of near misses, and personnel should be encouraged to report without fear of punishment. These reports should be analysed regularly to see if there are trends and to learn lessons.

If possible, national and even international networks should facilitate such reporting, because more can be learned from larger databases. ROSIS (Radiation Oncology Safety Information System) is an international voluntary reporting system that was set up on by ESTRO and is available on the web. It has been collecting data for several years and the value of the data continues to grow accordingly.

References

1. Toft B and Mascie-Taylor H 2005 Involuntary automaticity: A work-system induced risk to safe health care. *Health Services Management Research* **18**: 211–216.
2. IPEM (Institute of Physics and Engineering in Medicine) 2006 Balancing Costs and Benefits of Checking in Radiotherapy. IPEM, York.
3. IAEA (International Atomic Energy Agency) 2000 Lessons Learned from Accidental Exposures in Radiotherapy. Safety Report Series No. 17, Vienna.

CLINICAL APPLICATIONS OF 3-D CONFORMAL RADIOTHERAPY

RAYMOND MIRALBELL
Chef de service, Service de Radio-oncologie, Hôpitaux
Universitaires de Genève, Rue Micheli-du-Crest 24,
1211 Genève 14, Switzerland
Institut Oncòlogic Teknon, Barcelona-E
raymond.miralbell@hcuge.ch

Abstract Although a significant improvement in cancer cure (i.e. 20% increment) has been obtained in the last 2–3 decades, 30–40% of patients still fail locally after curative radiotherapy. In order to improve local tumor control rates with radiotherapy high doses to the tumor volume are frequently necessary. Three-dimensional conformal radiation therapy (3-D CRT) is used to denote a spectrum of radiation planning and delivery techniques that rely on three-dimensional imaging to define the target (tumor) and to distinguish it from normal tissues. Modern, high-precision radiotherapy (RT) techniques are needed in order to implement the goal of optimal tumor destruction delivering minimal dose to the non-target normal tissues. A better target definition is nowadays possible with contemporary imaging (computerized tomography, magnetic resonance imaging, and positron emission tomography) and image registration technology. A highly precise dose distributions can be obtained with optimal 3-D CRT treatment delivery techniques such as stereotactic RT, intensity modulated RT (IMRT), or protontherapy (the latter allowing for in-depth conformation). Patient daily set-up repositioning and internal organ immobilization systems are necessary before considering to undertake any of the above mentioned high-precision treatment approaches. Prostate cancer, brain tumors, and base of skull malignancies are among the sites most benefitting of dose escalation approaches. Nevertheless, a significant dose reduction to the normal tissues in the vicinity of the irradiated tumor also achievable with optimal 3-D CRT may also be a major issue in the treatment of pediatric tumors in order to preserve growth, normal development, and to reduce the risk of developing radiation induced diseases such as cancer or endocrinologic disorders.

Keywords: Conformal radiation therapy; computerized tomography; MRI; PET; 3-D CRT; IMRT; protontherapy; prostate cancer; brain tumors; pediatric tumors

1. Background

Radiation therapy (RT) has been used for the treatment of cancer for more than 100 years, and over the past 30 years its role has steadily increased. Radiobiology, medical physics, clinical oncology, and imaging/technology are the four pillars sustaining progress and development of contemporary radiation oncology.

1.1. RADIOBIOLOGY

Radiobiology studies the cell-killing effect of ionizing radiation and the mechanisms of cell-repair that are simultaneously triggered giving therefore the necessary scientific background for the use of RT to efficiently treat cancer. The cell killing effects of ionizing radiation are the consequence of their interaction with macromolecules, mostly DNA, RNA, and proteins. Such interactions may cause molecular breaks that trigger cell cycle regulatory actions (e.g., activation of cell cycle arrest genes) in order to repair the damaged macromolecules. DNA double strand breaks are, however, most difficult to repair and may lead to the death of the damaged cells at the time they attempt to divide (mitosis): the reproductive failure (Fig. 1). In addition, radiation may alternatively activate suicidal genes causing *apoptosis* or programmed cell death.

This later cell-killing mechanism is typical of lymphocytes, the most radiosensitive normal cell.

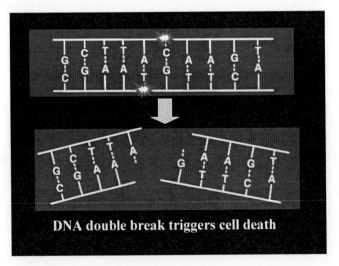

Figure 1. Example of DNA double break. See color section.

1.2. MEDICAL PHYSICS

Medical physics studies the different types and energies of radiation helping to optimize the ballistics for an optimal targeting of the tumours aimed to be destroyed. Since the 1960s, the radiation of choice in conventional RT has been high energy photons produced first in ^{60}Co teletherapy machines and now primarily by electron linear accelerators (linacs). Standard fractionation involves the delivery of daily fractions of 1.8–2 Gy, five times a week for a total dose ranging from 20 Gy for very radiosensitive tumours to 80 Gy or more for less radiosensitive ones.

1.3. CLINICAL ONCOLOGY

Clinical oncology studies the natural history of the different types of cancer, their growth, extension, diagnosis, and treatment alternatives. The most common cancers for women are breast, uterus, colon, lung and ovary, whereas for men it is lung, prostate, colon, head and neck, and bladder. In all these cancers, the large majority of patients present initially with localized disease. Curative RT can be given alone in some forms of cancers, like prostate cancer and early anal canal, and head and neck cancers. Though, more and more radiation is given now in conjunction with surgery, like breast cancer, brain tumours, soft tissue sarcoma etc., or in conjunction with systemic modalities (e.g. chemotherapy and hormone therapy or the new targeted treatments). Radiation oncologists, surgeons, medical oncologists, radiologists, and pathologists work together in "tumour boards" in order to establish the optimal multidisciplinary strategy in the diagnosis and treatment of individual patients suffering of cancer.

1.4. IMAGING AND TECHNOLOGY

Imaging and technology are the two engines pushing forward progress in precision offering the best possible tumour identification in order to obtain an optimal dose distribution differential between the tumour and the surrounding normal tissues. A better target definition is nowadays possible with contemporary imaging (computerized tomography, magnetic resonance imaging, and positron emission tomography) and image registration techniques (Fig. 2). Commercially available treatment planning systems offer fast and reliable dose distribution calculation methods in 3D once the target and the organs at risk have been defined during the virtual simulation procedure preceding the start of the treatment.

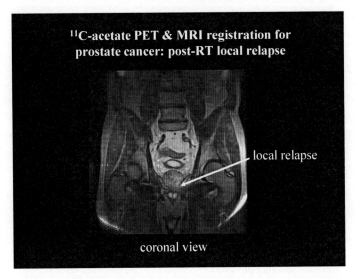

Figure 2. Example of prostate cancer. See color picture in Appendix I.

2. Present status of radiation oncology

After surgery, RT is nowadays the most effective anticancer treatment. Approximately 40% of the population in the western world will develop cancer and 60% of these patients will need RT either with a curative intent (60–70% of patients) or for symptom palliation (caused mostly by metastatic tumours such as pain, bleeding, or organ compression). A marked improvement in the outcome of patients diagnosed with cancer has been observed in the last 25–30 years. Early diagnosis and a multidisciplinary approach to treatment with an optimal association of surgery, RT, and chemotherapy explain, at least in part, such improvements. Indeed, the 5-year survival for adult patients treated in the US in the late 1970s and in the late 1990s was 43% and 64%, respectively, an improvement of 21% (roughly 1% per year).[1] A similar trend has been observed for cancer in children and adolescents for whom 5-year survival rates have improved from 58% to 77% for the same time intervals.[1]

Radiotherapy is not only efficient but is also the least expensive treatment method against cancer. A study published in 2003 assessing the cost of cancer treatment in Sweden estimated that RT was involved in 40% of cure rates but consumed only 5.6% of the total budget for oncology.[2] Recent progress in imaging and technology has significantly reduced side effects while succeeding to deliver potentially curative high doses. Thus, there is presently no existing alternative to RT for many cancers. Furthermore, the rate of cancer patients treated with RT has increased in the last decade. Indeed, in Geneva from 1994 to 2004, the rate of patients treated with RT has raised by a factor of 1.3. Cancer incidence will continue to rise mostly due to an increase in cancer cases due to the aging of the population and the growth of a population at risk such as the "baby boomers", soon entering the age of 50.

3. Still room for improvement in radiation oncology: 3D conformal RT optimization

Regardless of the improvement in cancer cure observed in the last 2–3 decades, roughly one third of patients still fail locally after curative RT. In order to improve local tumour control rates with RT the delivery of high doses to the tumour volume may be necessary.

Three-dimensional conformal radiation therapy (3-D CRT) first developed in the 1980s has been used to denote a spectrum of radiation planning and delivery techniques that rely on three-dimensional imaging to define the target (tumour) and to distinguish it from normal tissues. Few randomized studies have been carried out to prove that those technologies *"per se"* increased the local tumour control and decreased the rate of late complications, with some well-known exceptions like British trials of prostate carcinoma showing that 3-D conformal radiation was followed by significantly less complications than conventional radiation.[3] It is however commonly accepted in the radiation oncology community, that these technical advances have led, for the same "acceptable" rate of late complications (about 5%), to a higher rate of local control in a number of common cancers, like head and neck cancer, lung cancer, prostate cancer, breast cancer, some forms of brain tumours, gynaecological cancers etc.

Modern, high-precision RT techniques are needed in order to implement the goal of optimal tumour destruction delivering minimal dose to the non-target normal tissues. A highly precise dose distributions can be obtained with optimal 3-D CRT treatment delivery techniques such as stereotactic RT, intensity modulated RT (IMRT), or protontherapy (the latter allowing for in-depth conformation). Patient daily set-up repositioning and internal organ immobilization systems are absolutely necessary before undertaking any of the above mentioned high-precision treatment approaches.

4. Dose escalation to improve local tumour control

Prostate cancer, brain tumours, base of skull malignancies, and tumours growing in the soft tissues of the trunk are among the sites that potentially may most benefit of dose escalation approaches. The highest evidence in favour of dose escalation for prostate cancer has been obtained from several randomized trials undertaken in Canada, Europe, and the US with 3D-CRT or brachy-therapy.[4-6] Patients in these studies receiving doses of 75, 78 or 79.2 Gy (high-dose, study arms) presented with higher cure rates than those treated with 66 or 70 Gy (low-dose, control arms). In these studies, however, late toxicity was reported to be higher among patients treated in the high-dose arms suggesting a suboptimal 3D-conformation for normal tissues and the need of further progress in this field.

Pediatric tumours are usually radiosensitive and may not benefit of pure dose escalation approaches as in adults. Nevertheless, a significant dose reduction to the normal tissues in the vicinity of the irradiated tumour, also achievable with optimal 3-D CRT, may also be a major issue in the treatment of children and adolescents with cancer in order to preserve growth, normal development, and to reduce the risk of developing radiation induced diseases such as cancer or endocrinologic disorders.[7,8] Figure 3a–c present the case of an 8 year old child with a potentially curative tumour growing in the brain-stem (a pylocitic glioma) close to the sella turcica. A dosimetric optimization effort was under-taken in order to lower as much as possible the dose to the pituitary gland aiming to preserve the secretion of growth-hormone (GH). A sagittal view through the central target axis of the comparative plans between 3D-CRT and IMRT is presented. At last follow-up 6 years after IMRT the child is in remission and doing well with no growth or major endocrinologic side effects.

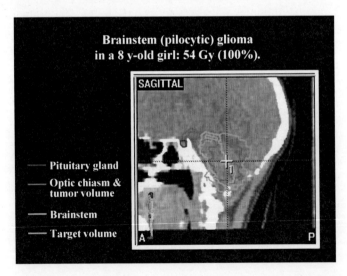

Figure 3a. Example of child brain-stem tumor. See color picture in Appendix I.

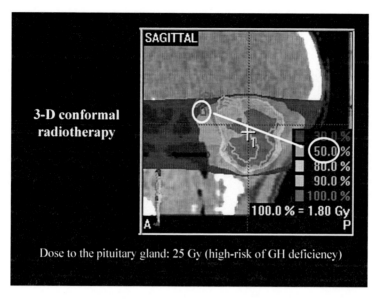

Figure 3b. Example of child brain-stem tumor. See color picture in Appendix I.

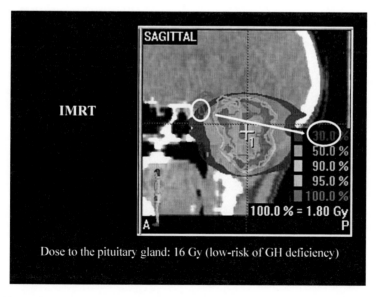

Figure 3c. Example of child brain-stem tumor. See color picture in Appendix I.

5. Summary and conclusions

High quality RT will continue to play an important role in the curative treatment of cancer in the future. It is likely that the number of cancer patients treated with RT will continue to increase in years to come. Better imaging, 3-D treatment planning, and precision will help to safely deliver higher doses for those tumours benefitting the most of dose escalation (in adults) and significantly reduce the dose to radiosensitive normal organs (in children). Thus, 3-D CRT optimization with IMRT (including novel technologies developed recently such as *Tomotherapy, Cyberknife, or RapidArc*) or proton beams may potentially improve local control rates while simultaneously reducing morbidity.

References

1. Jemal A, Clegg LX, Ward E, et al. Annual report to the nation on the status of cancer, 1975–2001, with a special feature regarding survival. Cancer 2004; 101:3–27.
2. Norlund A. Cost of radiotherapy. Acta Oncol 2003; 42:411–415.
3. Dearnaley DP, Khoo VS, Norman AR, et al. Comparison of radiation side effects of conformal and conventional radiotherapy in prostate cancer: a randomised trial. Lancet 1999; 353: 267–272.
4. Pollack A, Zagars GK, Starkschall G, et al. Prostate cancer radiation dose response: results of the M.D. Anderson Phase III randomized trial. Int J Radiat Oncol Biol Phys 2002; 53:1097–1105.
5. Zietman AL, DeSilvio ML, Slater JD, et al. Comparison of conventional-dose vs high-dose conformal radiation therapy in clinically localized adenocarcinoma of the prostate. A randomized controlled Trial. JAMA 2005; 294:1233–1239.
6. Sathya JR, Davis IR, Julian JA, et al. Randomized trial comparing Iridium implants plus esxternal-beam radiation therapy with external-beam radiation therapy alone in node-negative locally advanced cancer of the prostate. J Clin Oncol 2005; 23:1192–1199.
7. Miralbell R, Lomax A, Russo M. Potential role of proton therapy in the treatment of pediatric medulloblastoma/primitive neuroectodermal tumors: spinal theca irradiation. Int J Radiat Oncol Biol Phys 1997; 38:805–811.
8. Miralbell R, Lomax A, Cella L, Schneider U. Potential reduction of the incidence of radiation-induced second cancers by using proton beams in the treatment of pediatric tumors. Int J Radiat Oncol Biol Phys 2002; 54:824–829.

TREATMENT PLANNING: IMRT OPTIMIZATION – BASIC AND ADVANCED TECHNIQUES

SIMEON NILL
German Cancer Research Center, Research Program Imaging and Radiooncology, Department Medical Physics in Radiation Oncology e040, Im Neuenheimer Feld 280, D-69120 Heidelberg, Germany
s.nill@dkfz-heidelberg.de

Abstract This chapter will give an introduction on the different available intensity modulated radiation therapy (IMRT) optimization techniques.

Keywords: IMRT; optimization

1. Introduction

This chapter will give an introduction on the different available intensity modulated radiation therapy (IMRT) optimization techniques. The first part of this chapter explains the basic features of IMRT while the second part is intended as an outlook of what is currently under investigation by researchers all around the world. For a more comprehensive overview of IMRT the reader is referenced to the books of Professor Steve Webb[14, 15] or recently published review papers.[1]

2. The basics of IMRT optimization

The aim of every treatment technique is to achieve a conformal dose distribution to the target while reducing the dose to any organs at risk (OAR). The typical example used to demonstrate the features of IMRT is a horse shoe shaped target with an OAR in the middle (Fig. 1).

During conventional treatment planning all geometrical parameters are adapted by the treatment planner in such a way as to obtain the best possible treatment plan. Typical free parameters are the energy, the isocenter position inside the patient, the beam directions, the leaf positions and the weighting factor of the fluence. After these parameters are fixed the treatment planning system (TPS) starts the dose calculation and the resulting dose distribution is analysed. If the result is not sufficient to achieve the predefined constraints the user manually adapts the free parameters and restarts the dose calculation. This workflow is called forward treatment planning. The main difference between conventional therapy and IMRT is that the fluence distribution from one beam direction is not uniform any more but the 2D fluence distribution is now a function of the position x and y in the beam's eye view coordinate system. On the left side of Fig. 1 the yellow arrows for each beam have the same length at each position inside the treatment field whereas on the right side these lengths of the arrows are modulated. Due to this modulation the resulting 3D dose distribution for IMRT improves the sparing of the OAR (green). This chapter will briefly explain how this fluence modulation is calculated based on the planning constraints set by the treatment planner. Since for the IMRT planning process the user defines the dose distribution that he would like to obtain and the system then calculates the treatment parameters using a mathematical optimization algorithm this process is called inverse treatment planning.

The standard loop of an inverse treatment planning system is outlined in Fig. 2 and consists of the following boxes:

1. Treatment Parameters
2. Dose Calculation
3. Objective Function based on clinical experience
4. Optimization algorithm

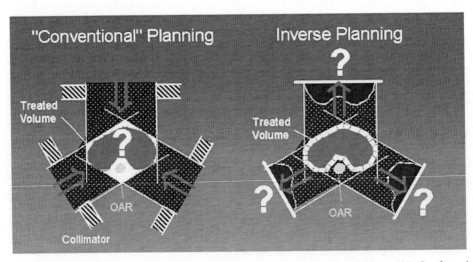

Figure 1. Principle of inverse planning. (Images taken from: Schlegel and Mahr, "3D Conformal Radiation Therapy: multimedia introduction to methods and techniques", 2nd revised and enhanced edition, Springer Verlag.)

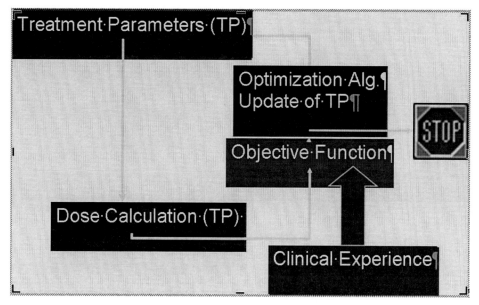

Figure 2. Standard optimization loop.

The next section explains in more detail the content of each of the boxes.

2.1. TREATMENT PARAMETERS

As already outlined in the introduction there are multiple parameters which a treatment planner can change to obtain the best treatment plan. Table 1 contains a selection of these parameters and some further explanations. Each of these parameters can be optimized and are therefore called free parameters. The more parameters that are optimized the more degrees of freedom the optimizer has to find the optimal treatment plans.

2.2. DOSE CALCULATION

The dose calculation engine applied during the optimization strongly depends on the TPS used. Some systems are using an improved pencil beam algorithm while others are using a superposition algorithm or even Monte Carlo based dose calculation engines. There are some tumor sites where more advanced dose calculation algorithm than pencil beam really should be used (see second part of this chapter).[5,9]

TABLE 1. Possible treatment parameters to optimize.

Treatment parameters	Comments
Intensity profiles (fluence)	These are the basic parameters which are either directly or indirectly optimized by all inverse TPS systems.
Beam angles (gantry, couch angle) Number of beams	Some systems are requiring these geometrical beam settings as input parameters while other systems are capable of optimizing these parameters. Beam angle optimization is still an ongoing research topic. Currently for most cases 5–9 equidistant spaced coplanar beams are used.
Energy	For photon IMRT the energy is most often fixed to a preselected value for the whole treatment plan while for IMPT (intensity modulated particle therapy) the energy is a free parameter which is optimized by the TPS.
Type of radiation (photons, electrons, protons, carbons, …)	Currently most TPS require the user to predefine the radiation type before the optimization. Combined optimization or even computer bases selection of the radiation type is an ongoing research topic.

2.3. OBJECTIVE FUNCTIONS BASED ON CLINICAL EXPERIENCE

To perform an automatic optimization of the treatment parameters the system needs a method to rank the different calculated treatment plans. Therefore clinical experience must be transformed into mathematical equations which are then used by the optimization algorithms to derive the optimal treatment plan. The outlined task is to correlate physical information (3D dose distribution) to patient outcome in terms of tumor control and radiation induced complications. There a two approaches to this task. The first possibility is to used physical dose thresholds based on the calculated dose volume histograms (DVH)s where as the second more ambitious approach tries to correlate the physical parameters directly to biological parameters such as tumor control probability (TCP) or normal tissue complication probability (NTCP). Due to the still existing uncertainties of the biological parameters most systems are still using the traditional DVH constraints as input for the objective functions. The objective function converts the DVHs into a mathematical norm to compare different treatment plans. The most common used objective function is the quadratic objective function which is the square measure of constraint violations. This function is mainly chosen for simplicity reasons.

The basic dose constraints have to be entered into the graphical user interface of the inverse TPS. In addition to the dose thresholds the user has to assign a penalty factor. This relative penalty factor controls the importance of the individual dose constraints.

2.4. OPTIMIZATION ALGORITHMS

To optimize the treatment parameters different optimization algorithms are available. The algorithms currently most often used can be grouped into two categories.

- Deterministic gradient methods
- Stochastic methods

Both of these methods have their advantages and disadvantages. Deterministic gradient methods are normally very fast but can become trapped in local minima. Therefore this group is most often applied when there are no local minima or the local minima is almost as good as the global minimum as it is the case for the quadratic objective function in combination with DVH constraints. Stochastic methods like simulated annealing or genetic algorithms can be applied to any arbitrary objective function but are slower than gradient methods but tend to find the global minimum. The best way is always to select the most suited optimization algorithm for a specific objective function.

2.5. WORKFLOW

The typical IMRT workflow starts after the target and the OAR were delineated. The first step is to assigning the constraint thresholds and penalty parameters to the targets and the OARS (see Fig. 3). The computer then optimizes the treatment parameters and finally calculates the 3D dose distribution. This distribution is then evaluated by the treatment planner. If there are modifications necessary the treatment planner adapts the constraint parameters (Fig. 3) and the optimization is rerun.

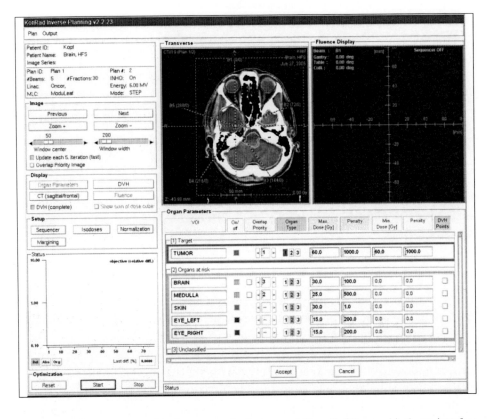

Figure 3. Inverse treatment planning system. In the lower right part of the graphical user interface the dose base constraints can be entered. (Images taken from: Schlegel and Mahr, "3D Conformal Radiation Therapy: multimedia introduction to methods and techniques", 2nd revised and enhanced edition, Springer Verlag.)

2.6. EXAMPLE

Figure 4 shows the graphical user interface of the treatment planning system KonRad. The dose distribution shows a typical IMRT dose distribution for an intra cranial tumor.

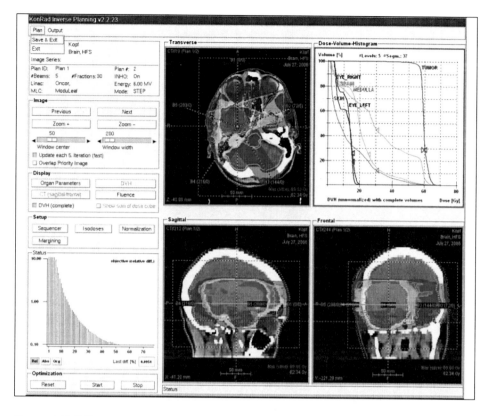

Figure 4. IMRT treatment system with a typical IMRT dose distribution. (Images taken from: Schlegel and Mahr: "3D Conformal Radiation Therapy: multimedia introduction to methods and techniques", 2nd revised and enhanced edition, Springer Verlag.)

3.   Advanced IMRT optimization

Most of the topics covered in the second part of this chapter will only be briefly described since most of them are still being actively investigated by researchers.

There are numerous sections which could be improved in the current IMRT workflow but only the following three issues are being discussed in this section:

1. Beam angle optimization
2. Advanced dose calculations
3. Inclusion of hardware constraints into the optimization

3.1. BEAM ANGLE OPTIMIZATION

Most inverse TPS systems are currently requesting that the number of beams
and their direction of incidence are set to a fixed value by the treatment planner.
The idea of beam angle optimization is not a new topic for IMRT since it was
also pursued for conventional treatment techniques. The current assumption/
belief is that five to nine equidistant spaced beams are providing enough degrees
of freedom to ensure optimal plans. This is probably correct for simple targets
like prostate tumors but for highly complex cases like head-and-neck tumors
there is still room for improvement. Including the beam angles into the optimization increases the complexity and also, depending on the integration of the
beam angles into the optimization algorithm, can also lead to non-convex objective
functions.[7] For these kinds of objective functions stochastic algorithms are most
often required to find the best solution and therefore the computational time is
significantly increased. The resulting dose distributions are often of higher
quality but sometimes the differences are small and it is not clear whether they
are clinical significant.

3.2. ADVANCED DOSE CALCULATIONS

Currently most treatment planning systems are using an improved pencil beam
dose calculation algorithm for IMRT optimization. For some indications where
there is less tissue-inhomogeneities the accuracy of the dose calculation engine
is acceptable with the exception of complex head-and-neck cases or especially
for small tumors inside the lung current algorithms. Scholz et al.[9, 10] and others[5]
have showed that there are significant differences in terms of systematic and
convergence errors for these kinds of tumors. The systematic error indicates the
difference of the two final dose distributions when the fluence distribution
(obtained employing dose engine A) is recalculated once with dose engine A
and dose engine B (equation 1). The convergence error on the other hand is
defined as the dose difference obtained when different dose engines are used
during the optimization step but always the same algorithm is used for the final
recalculation of the 3D dose distribution (equation 2).

$$\Delta_s = d_{A-opt}^{A-recalc} - d_{A-opt}^{B-recalc} \tag{1}$$

$$\Delta_c = d_{A-opt}^{B-recalc} - d_{B-opt}^{B-recalc} \tag{2}$$

Figure 5 shows the two different dose distributions used to evaluate the systematic error for a small lung tumor. On the left side the pencil beam algorithm (PB)
while on the right side a superposition algorithm (S) was used. It can clearly be
seen that the 90% isodose does not cover the CTV anymore.

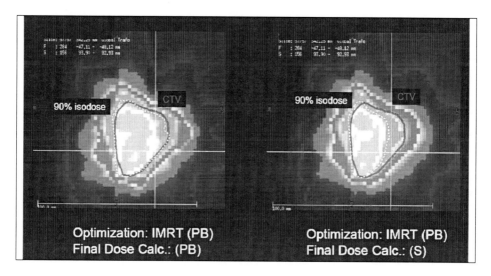

Figure 5. Comparison of two different dose calculation engines for a small lung tumor.[9]

The same comparison could also be performed with even more advanced dose calculation engines such as Monte Carlo.[5, 9] If the superposition algorithm is already used during the optimization it was shown by Scholz[9] that for targets in the head-and-neck area the systematic difference between Monte Carlo and Superposition algorithm is small but for regions of high density gradients like the lung the Monte Carlo dose algorithm if employed already during the optimization stage lead to the best results.

3.3. INCLUSION OF HARDWARE CONSTRAINTS INTO THE OPTIMIZATION

The majority of existing treatment planning systems is using a two step approach during the optimization loop to calculate the leaf sequences. After the optimization algorithm calculated the 2D fluence a sequencer is applied to convert the fluence maps into either dynamic or static step and shoot leaf sequences. Based on these sequences the final dose distribution is then calculated. During the conversion step, the hardware constraints of the multi leaf collimator are taken into account and therefore there is a difference between the "optimal" fluence and the "deliverable" fluence. Depending on the delivery technique and the number of stratification levels the final dose distribution might look significantly different than the optimized dose distribution. One solution to overcome this problem is to use aperture based optimization techniques which have the advantages that no leaf sequencing is necessary and the MLC hardware constraints are automatically taken into account.[11] There are different aperture based techniques currently available. The first group is called "contour-based". For this optimization technique the algorithm creates a fixed number of MLC shapes which are either anatomy[2] or isodose based[4] and then only optimizes the weight of the segments until the optimization goal is reached. The advantages of this algorithm are that accurate dose calculation engines could easily already be used during the optimization process and that only a few intuitive MLC shapes are generated. The major disadvantages are that no general shape generation algorithm exists which suits all indications and that the leaf positions are not varied during the optimization process.

The second group of algorithms is called direct aperture optimization (DAO).[12] The main difference to the first group is that now not only the weight of the segments but also each leaf position is optimized during the optimization loop. The user only needs to specify the number of shapes per beam in addition to the already known target and OAR constraints. Most algorithms are using stochastic algorithms like simulated annealing to calculate the optimal leaf positions. Using the DAO technique it is possible to obtain treatment plans with reduced monitor units, reduced number of segments which then leads to a reduced delivery time.

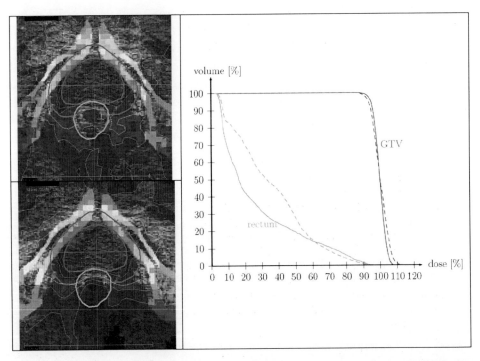

Figure 6. Comparison between a 6 MV X field IMRT plan (top row) with an AMCBT plan (bottom row[16]). The solid lines in the DVH represent the AMCBT plan. (Courtesy of Silke Ulrich.) (See color picture in Appendix I).

Compared to the normal sequencing process the user can control the plan complexity by simply reducing or increasing the number of shapes per beam. A minor drawback is the increased optimization time which will be less of importance in the future due to continuously increasing computer power.

The latest advances in radiation therapy using the DAO approach are dynamic rotational delivery techniques.[6,16] Compared to the already established rotational techniques (IMAT) the newest techniques not only dynamically vary the leaf positions and speed dependent on the gantry angle but also the dose rate and even the gantry speed are modulated. Using this technique, highly conformal IMRT plans can be realized within one or two gantry rotation leading to a possible reduction of the treatment time (see Fig. 6). Due to the higher complexity of the dose delivery technique an increased QA effort for the patient and machine QA is expected.

4. Conclusion

IMRT offers the possibility to obtain highly conformal dose distributions and is applied routinely to multiple clinical indications. Due to the high conformality of the dose distributions IMRT should only be applied together with image guidance.[3]

References

1. Bortfeld T., IMRT: a review and preview, Phys. Med. Biol. 2006 51(13):R363–379.
2. De Meerleer et al., Radiotherapy of prostate cancer with or without intensity modulated beams: a planning comparison. Int J. Oncol. Biol. Phys. 2000 47:639–648.
3. Image-Guided IMRT T., Bortfeld R., Schmidt-Ulrich W., de Neve D.E., Wazer (eds), Springer-Verlag Berlin Heidelberg Germany; ISBN 354020511x, 460.
4. Kestin et al., Intensity modulation to improve dose uniformity with tangential breast radiotherapy: initial clinical experience. Int. J. Oncol. Biol. Phys. 2000 48:1559–1568.
5. Mihaylov I.B., Siebers J.V., Evaluation of dose prediction errors and optimization convergence errors of deliverable-based head-and-neck IMT plans computed with a superposition/convolution dose algorithm. Med. Phys. 2008 35(8):3722–3727.
6. Otto K., Volumetric modulated arc therapy: IMRT in a single gantry arc. Med. Phys. 2008 35:310–317.
7. Pugachev A. et al., Int. J. Oncol. Biol. Phys. 2001 50(2):551–560.
8. Schlegel W. and Mahr A., 3D Conformal Radiation Therapy: multimedia introduction to methods and techniques, 2nd revised and enhanced edition, Springer Verlag Berlin 2007.
9. Scholz C., Development and evaluation of advance dose calculations for modern radiation therapy, Ph.d. thesis, University of Heidelberg (2004).

10. Scholz C. et al., Comparison of IMRT optimization based on a pencil beam and a super-position algorithm. Med. Phys. 2003 30(7):1909–1913.

11. Shepard D.M. et al., Aperture based inverse planning, in: Palta J.R., Mackie T.R. (eds) Intensity modulated radiation therapy: the state of the art. Medical Physics Publishing, Madison, WI, 2003 pp 115–137.

12. Shepard D.M. et al., Direct aperture optimization: a turnkey solution for step-and-shoot IMRT. Med. Phys. 2002 29(6):1007–1018.

13. Siebers J. et al., Med. Phys. 2002 29(6):952–959.

14. Webb S. 1997, The Physics of conformal radiotherapy: advances in technology, Institute of Physics Publishing, Bristol/Philadelphia, PA.

15. Webb S. 2001, Contemporary IMRT: developing Physics and clinical implementation, Institute of Physics Publishing, Bristol/Philadelphia, PA.

16. Ulrich S. et al., Development of an optimization concept for arc-modulated cone beam therapy. PMB 2007 52:4099–4119.

TREATMENT PLANNING: STEREOTACTIC TREATMENT TECHNIQUES

SIMEON NILL
*German Cancer Research Center, Research Program Imaging
and Radiooncology, Department of Medical Physics in Radiation
Oncology e040, Im Neuenheimer Feld 280, D-69120
Heidelberg, Germany*
s.nill@dkfz-heidelberg.de

Abstract This chapter will give an introduction to the use of stereotactic treatment techniques in radiation therapy.

Keywords: Treatment planning; stereotaxy

1. Introduction

This chapter will give an introduction to the use of stereotactic treatment techniques in radiation therapy and is divided into three sections. More detailed explanations can be found in the literature or on the MMCD published by Springer.[6]

The basic idea behind any stereotactic guided treatments is to put the position of a hidden and possibly small target in correlation to an external coordinate system during the whole treatment chain. It is therefore of paramount importance to preserve this correlation once it is established and therefore most often mechanical fixation systems are used. The basic concept of stereotactic system for intracerebral surgery was introduced by Lars Leksell[3, 4] in 1948 but the adaptation to stereotactic treatments of the whole body was only achieved recently (mid 1990).[1]

2. Basis stereotactic tasks

The basic tasks for every stereotactic intervention or treatment can always be described as:

1. Patient fixation
2. Target Localization and treatment planning
3. Target positioning and intervention/treatment

2.1. PATIENT FIXATION

For the fixation of the patient different fixation solutions are available and it is therefore not possible to list them all. The interested reader is therefore referred to the internet where numerous vendors are offering their solutions. Since no fixation system can be used for all tumor locations these systems are split into two groups. The first group is specialized on the fixation of the patient's head (intra-cranial treatments) where the second group focuses on the treatment of tumours in the whole body (extra-cranial). Independent of the localization the fixation devices are divided into invasive and non invasive systems. For stereo-tactic radiosurgery (SRS) (single fraction treatment) a very high precision is required and therefore invasive fixation systems are used whereas for fractionated stereotactic radiotherapy (SRT) non invasive fixation devices are applied. Despite the higher localization requirement for SRS this is mostly due to the fact that for a whole treatment course of 20–30 fractions no patient is willing to live with an invasive fixation system. Another import issue is the basic material of such a fixation system. If the system is only to be used during computed tomography the basic ring (Fig. 1a) can be made of metal but if multi-modality images are necessary to precisely locate the tumour MR or even PET compatible materials must be used. In Fig. 1b the base ring is therefore made out of wood.

2.2. PATIENT/TARGET LOCALIZATION

After the correlation between the patient and the fixation system is established the next task is to locate the target volume. Therefore multiple imaging modalities exist. For basic treatment planning at least a computed tomotherapy of the patient with the localizers (see Fig. 1b) attached must be performed. Other possible modalities are 3D MR/PET images or especially for treatments of intra-cranial lesions an angiogram is often performed. For each of these systems individual localizers needed to be attached to the base ring.

The following paragraph describes the localizer used at the DKFZ (German Cancer Research Center Heidelberg). The basic shape of the localizer can be compared to the to the letter V which is rotated by 90° with three wires. The central wire is parallel to the patient axis. The angle between the outer two wires is selected in such a way that for each position the distance between the outer two markers equals the stereotactic slice position. The material of these wires is tungsten or steel for the CT localizer and for the MR system the tubes are filled with Gd-DTPA. Figure 1c shows a transversal slice of a CT images. On each side of the pictures the three markers can be clearly seen. The intersection of the lines connecting the opposing central markers indicates the origin in the x and y direction. Once the coordinate system is established for each indication the delineated structures in one dataset can easily be displayed on any other 3D dataset since they share a common coordinate system. The next step is to perform 3D treatment planning. Since SRS requires special accessories like spherical cones or micro-multi-leaf collimators often separate treatment planning system are used. These systems are often sold together with the necessary equipment to perform SRS.

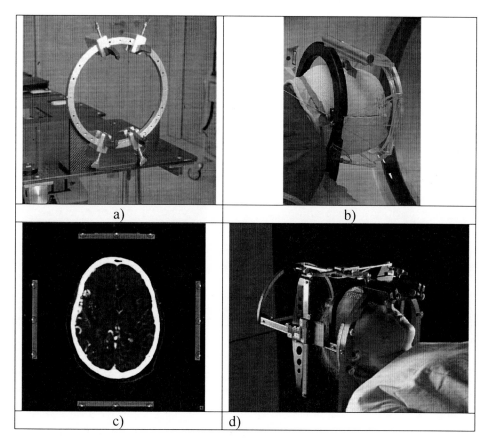

Figure 1. (a) Invasive fixation system. (b) Non-invasive scotch cask mask with attached localizers. (c) Transversal CT slice. (d) Target positioning system. (Images taken from: Schlegel and Mahr, "3D Conformal Radiation Therapy: multimedia introduction to methods and techniques", 2nd revised and enhanced edition, Springer Verlag.)

2.3. TARGET POSITIONING

After the treatment plan is created the treatment planning system calculates the stereotactic coordinates for each isocenter. The patient is positioned on the treatment couch and now the task is to find the stereotactic target point and to align this point with the room lasers. Therefore a target positioning system is mounted on the base ring of the stereotactic system. This target positioning system at the DKFZ is equipped with electronic callipers which are now moved along the main axis to display the calculated stereotactic coordinates (Fig. 1d). After the numbers are double checked by another operator the couch is moved to align the stereotactic target point with the room lasers. If everything is setup correctly the treatment can begin. For this procedure the room lasers have to be precisely aligned with the radiological isocenter of the linear accelerator. This can be achieved for example by the Winston Lutz test.[5]

Using the three outlined steps it is possible to treat a hidden target with a very high precision but there are several points which must be taken into account. Even with a fixation system there could be some residual errors or even some systematic offsets. Therefore the whole treatment chain has to be carefully validated. One possible source of error can be the fixation system itself. If a patient with an invasive fixation system unintendedly applies some force on the fixation system during the time between the imaging session and the treatment the once established correlation could be incorrect. Therefore it is very important to perform a validation prior to the start of the treatment with the patient in the treatment position. This could either be some MV portal images or even a 3D cone beam CT.[2]

3. Modern stereotactic treatment techniques

As stated in the introduction of this chapter the stereotactic fixation was introduced a long time ago. Today this technique is most often used for the application of stereotactic radiosurgery (SRS) or fractionated stereotactic radiotherapy (SRT). The basic concept is to combine a steep dose gradient with a precise localization of the target. SRS is normally applied for single fractioned high dose therapies of small lesions where SRT focus on multiple fractions and larger lesions.

One commercially available system with the focus on SRS is the GammaKnife™ (Elekta). It uses 201 cobalt sources positioned in a hemisphere and aimed through round collimators to a common focal point.

Only after the linac manufacturers obtained a high mechanical precision of the gantry and the couch rotation was it possible to use a conventional linac equipped with specialised accessories like round collimators or micro multileaf collimator to perform SRS (3). The first treatment techniques for SRS using a linear accelerator were multiple spherical arcs. With this technique it is possible to obtain a good target coverage and conformity for small spherical lesions but for irregularly shaped target multiple isocenters must be used. Still the dose distribution is not as good as for spherical lesions and the treatment time is prolonged sometimes over 1 h. Therefore multiple non-coplanar fixed field irradiation techniques with up to 14 irregular shaped beams were used to obtain an improved dose calculation (Fig. 2).

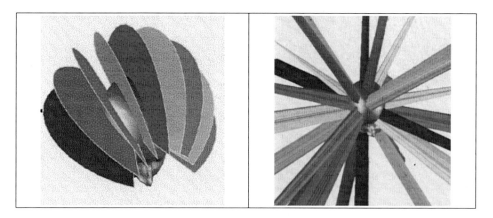

Figure 2. Left side: Multiple arc rotations with one single isocenter. Right side: 14 individual shape fields (multi field irradiation).

Nowadays more modern techniques like intensity modulated radiation therapy (IMRT) are capable of delivering very conformal dose distributions even for concave shaped targets. Possible IMRT techniques are IMRT with compensator or with integrated or add-on multi leaf collimators (MLC).

Therefore the concept of stereotactic localization is often still applied together with IMRT but there are other possible localization techniques available which are currently challenging stereotaxy. Today more and more linear accelerators are equipped with the possibility to acquire 3D images of the patient in the treatment position[2] (Fig. 3). Using intelligent and precise image registration algorithms it is possible to align the target with the isocenter without using any stereotactic localisation system during the acquisition of the treatment planning CT.

The use of image guidance for patient positioning might render the stereotactic localization technique unnecessary in the future but there is still the problem of immobilization. With the stereotactic mask or ring the patient could not or only slightly move. Without these devices other methods to monitor patient motion during the treatment like infrared cameras or laser surface scanner are required and are integrated into the clinical workflow.

112 S. NILL

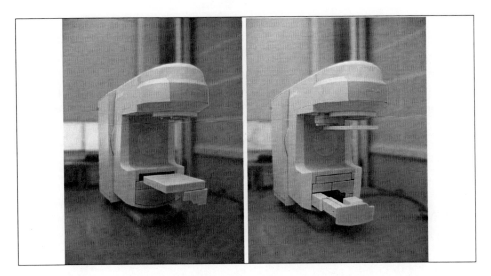

Figure 3. Modern linear accelerator with MV and kV imaging capabilities. (Images taken from Schlegel and Mahr: "3D Conformal Radiation Therapy: multimedia introduction to methods and techniques", 2nd revised and enhanced edition, Springer Verlag.)

4. Conclusion

Stereotactic methods provide a fixed correlation between the patient and the stereotactic system. Using this method high precision radiotherapy treatments are possible. Modern image guided techniques might make the use of these systems not necessary any more in the future.

References

1. Hartmann GH, Bauer-Kirpes B, Serago CF, Lorenz WJ. Precision and accuracy of stereotactic convergent beam irradiations from a linear accelerator. Int J Radiat Oncol Biol Phys. 1994 Jan 15; 28(2):481–92.
2. Jaffray DA. Emergent technologies for 3-dimensional image-guided radiation delivery. Semin Radiat Oncol. 2005 Jul; 15(3):208–16.
3. Leksell DG. Stereotactic radiosurgery. Present status and future trends. Neurol Res. 1987 Jun; 9(2):60–8.
4. Leksell L. The stereotaxic method and radiosurgery of the brain. Acta Chir Scand. 1951 Dec 13; 102(4):316–9.
5. Lutz W, Winston KR, Maleki N. A system for stereotactic radiosurgery with a linear accelerator. Int J Radiat Oncol Biol Phys. 1988 Feb; 14(2):373–81.
6. Schlegel and Mahr. 3D Conformal Radiation Therapy: A multimedia introduction to methods and techniques. 2nd revised and enhanced edition, 2007, Springer Verlag Berlin.

IMAGE GUIDED RADIOTHERAPY

UWE OELFKE
DKFZ, Heidelberg, Im Neuenheimer Feld 280,
69120 Heidelberg, Germany
u.oelfke@dkfz-heidelberg.de

Abstract Image guided radiation therapy (IGRT) is currently one of the most active research fields in medical physics. The recent development of high precision dose delivery techniques with high energy photon and hadron beams can only be fully exploited if we confidently know the shape and location of radiation targets and the organs at risk at the time of the treatment. Most of the time, anatomical images of the patient were only acquired for the purpose of diagnosis and treatment planning usually days before the treatment planning process, i.e., the patient anatomy was assumed to be static in time. As a result, all time dependent variations of relevant anatomical structures, like shifts and deformations of tissues in time, for instance also caused by the radiation treatment, were not accounted for and therefore are a significant source of potential treatment errors. We briefly describe the basic concepts and technology employed in current IGRT practice.

Keywords: Image guided radiation therapy; adaptive radiation therapy; cone beam CT

1. Introduction

Standard clinical practice in radiation therapy, even for most sophisticated treatment techniques like intensity modulated radiation therapy (IMRT), is usually based on a set of assumptions and concepts that aim for a safe and reliable treatment. These assumptions specifically deal with geometrical uncertainties of the irradiation setup and the patient's anatomy during the course of treatment.

With respect to the most important input data for treatment planning, mostly anatomical CT and MRI images taken at least days if not weeks prior to treatment, it is for instance anticipated that they are static in time. Only minor deviations should occur between image acquisition and patient treatment.

Y. Lemoigne and A. Caner (eds.), *Radiotherapy and Brachytherapy*,
© Springer Science + Business Media B.V. 2009

Furthermore, we assume that geometrical uncertainties caused by patient setup, dose delivery and organ motion of the radiation target can be reliably accounted for by appropriate CTV to PTV safety margins. Usually, no specific measures are taken for the respective geometrical misalignments of organs at risk and radiosensitive anatomical structures.

Unfortunately, the magnitude of these assumptions could only be estimated and hardly be verified by quantitative measurements. Furthermore, it is well known, that other long term trends like tissue responses to radiation or weight loss of a patient will modify the irradiation geometry from the one employed for treatment planning. All these effects accumulated, therefore can lead to a considerable uncertainty whether the optimized planned treatment was actually delivered to a patient.

To resolve this problem of radiation therapy additional x-ray imaging of the patient in treatment position, either directly prior or during treatment, was developed in the last decade.[1] These imaging devices provide quantitative data of the actual irradiation geometry which forms the basis for image-guided radiation therapy (IGRT). IGRT aims to detect and correct for all time dependant changes of the irradiation geometry encountered during a course of radiotherapy. Specifically, the following sources of geometrical irradiation errors can potentially be addressed by IGRT:

- Patient and radiation target setup errors.
- Inter- and intra-fraction organ motion.
- Short term organ deformations.
- Long term anatomical changes (weight-loss or -gain, tissue swelling).
- In the following we will give a brief overview about some of the concepts and technologies related IGRT and adaptive radiotherapy (ART).

2. The concept of adaptive radiation therapy

The general set of strategies for an adaptation of treatment parameters based on images taken close or during treatment is called adaptive radiation therapy (ART).[2] In the following we will first outline some general features of an ART system and describe its basic components.

Depending on the anticipated rate of anatomical changes, as for instance caused by different types of organ motion, various levels of adaptive radiation therapy can be imagined. For inter-fraction errors like patient setup errors, on-line and off-line protocols of ART can be performed.[3, 4] A first level of ART can be achieved by the acquisition of several CT scans of the patient in treatment position prior to the beginning of the treatment. A statistical analysis of these imaging

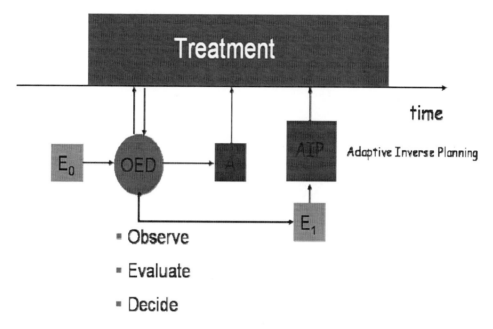

Figure 1. The general concept of ART: Within the time of treatment the irradiation geometry is observed by a number of imaging procedures. The information provided by these images has to be evaluated by a comparison with an initial expectation E_0., for instance the images employed for treatment planning. Finally, a decision about the adaptation process has to be made. This can for instance be a direct, online adaptation (A) of treatment parameters if the observation can be easily interpreted. An off-line strategy like adaptive inverse planning may be adopted if further continuing accumulation of evidence E_1 is required.

data can then be used for the definition of an adaptive protocol. Another strategy would be to adapt the treatment parameters directly prior to any fraction of the treatment based on a daily CT of the patient. The most sophisticated strategy would be a dynamic refinement of the treatment plan during treatment based on real-time imaging data (Fig. 1).

3. IGRT imaging equipment

3.1. IN ROOM CT TECHNIQUES

The most common imaging technique is the integration of CT imaging into the treatment machine. This can be achieved by the methods of kV- and MV-cone beam or fan-beam CTs.[5]

3.1.1. *In room kV-CTs*

There are two in-room imaging techniques based on x-ray imaging with conventional x-ray tubes in the kV-energy regime.

3.1.1.a. CT on rails

The first IGRT solution, provided by Siemens/OCS, is the integration of a CT scanner 'on rails' in the treatment room.[6] The CT scanner and the treatment machine share the treatment couch, i.e. the couch can be either rotated or shifted from the treatment position into an imaging position where the CT-on rails can take a CT image. The utilization of these images for the adaptation of the patient setup requires an accurate calibration of the treatment and imaging coordinate frames of reference. The main advantage of this system is its excellent imaging quality being achieved at dose levels known from diagnostic imaging.

3.1.1.b. kV cone beam CT

The majority of IGRT procedures are currently performed with linac-integrated cone beam CTs.[7] A kV x-ray source and an oppositely mounted flat panel detector are integrated with the gantry of the linac. With this approach two imaging geometries can be realized: (i) the imaging beam is aligned perpendicular to the axis of the treatment beam (90° solution, VARIAN, ELEKTA) and (ii) the sources of the imaging beam and the treatment be are mounted along the same axis (SIEMENS, OCS).[8] This 180° solution has the additional advantage that the treatment field and the relevant patient anatomy can be also monitored during treatment. Currently, only the 90° solutions are commercially available.

The main advantage of the linac-integrated CT-imaging is that the patient setup at the linac is accomplished without any further delay or unnecessary patient movement. One has to be aware, that the restriction of the gantry rotation speed of the linac limits the image acquisition to roughly 30 s for a short scan and 60 s for full scan. Furthermore, for short scans the field of view is usually restricted to 25–27 cm. The vendors usually provide an extended field of view options for a complete gantry rotation. In comparison to the fan-beam CT of the 'CT-on rails' approach cone beam CTs naturally cannot provide the same excellent image quality.[9] The main factors for that are the enhanced scatter contribution and the limited quantum detection efficiency of the employed flat panel detectors.

3.1.2. *In-room MV-CTs*

MV-CTs are attractive for IGRT imaging because this approach does not require any new radiation source. The MV-treatment beam itself is directly employed for the acquisition of the CT projections. As for the kV-imaging there are currently two different types of devices commercially available. The main disadvantage of the MV approach is the naturally reduced soft-tissue contrast in comparison to standard kV-x-ray spectra.

3.1.2.a. MV-cone beam imaging

The MV-cone beam imaging approach is currently commercially available by Siemens/OCS.[10] The image projections are detected with a flat-panel electronic portal imager optimized for its use with MV beam quality. Scatter and the modest soft-tissue contrast of MV-beams pose limitations to the achievable image quality.

3.1.2.b. MV-fan beam imaging

This imaging method is provided by the concept of tomotherapy (Tomotherapy, Inc.).[11] This technology provides a helical scanning pattern while the patient is moved with the couch through the gantry. Because of the fan-beam characteristics the scatter contribution to the images is reduced when compared to the MV-cone beam approach. Furthermore, two additional measures were taken to improve the image quality: (i) reduction of the average energy of the bremsstrahlungs-spectrum and (ii) the use of high CT detector with high QDE.

3.2. THE CYBERKNIFE

For the Cyberknife system dynamic dose delivery is accomplished by a 6 MV linac mounted to an industrial robot, that allows to irradiate target volumes from hundreds to thousands of selected beam directions. Image guidance is provided by two kV x-ray sources mounted to the ceiling of the treatment room.[12] With these x ray sources fluoroscopic images of the patient's anatomy or implanted radio-opaque markers during treatment are recoded by two floor-integrated flat-panel imagers. Furthermore, a set of infra-red markers distributed on the patient surface can be continuously detected by a camera system. A sophisticated correlation procedure between external surface movements, detected by the cameras, and x-ray images revealing the internal patient anatomy is employed to dynamically adapt the robotic dose delivery system such that organ motion can be automatically compensated for.

4. Correction of inter- and intra-fraction treatment errors

4.1. CORRECTION OF INTER-FRACTION TREATMENT ERRORS

The most common procedure for the correction of inter-fraction errors is an automated patient setup based on CT-images of the relevant anatomy in treatment position acquired directly prior to treatment. The images, originating either from kV- or MV-cone beam or fan beam CTs, are then compared to the diagnostic CT images available for treatment planning. A successful software registration of the two sets of imaging data usually results in new coordinates for the patient setup. Mostly, this is accounted for by a translation of the patient by an automatic shift of the treatment couch. This simple IGRT procedure, whose basic workflow is briefly sketched in Fig. 2, usually requires an extra time span of 5–10 min for each treatment.

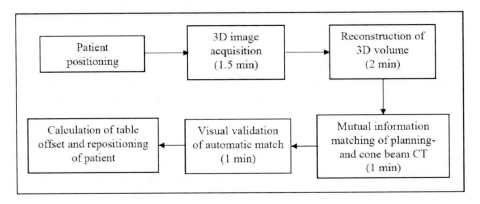

Figure 2. Simplified workflow for automated patient positioning via 3D-CT images taken directly prior to treatment.[13]

4.2. CORRECTION OF INTRA-FRACTION ORGAN MOTION

This section will focus on the correction of intra-fraction treatment errors, i.e., treatment errors that occur during the treatment often caused by breathing induced organ motion. The respective imaging and patient monitoring approaches as well as treatment strategies like 'gating' and 'tracking' will be introduced. These concepts are referred to as online strategies for treatment adaptation.

4.2.1. *Motion induced imaging artifacts*

Organ motion in the time scale of seconds or a few minutes can significantly diminish the quality of well optimized dose distribution. The most obvious effect is the well known blurring of dose gradients if the organ movement is directed perpendicular to the dose gradient. A similar effect, mostly observed for breathing induced organ motion, already surfaces at the point of image acquisition, where for instance CT projections taken for a lung tumor in motion do not necessarily all belong to the same breathing phase. An example for the resulting severe artifacts is shown in Fig. 3. Consequently, modern 4D-CT

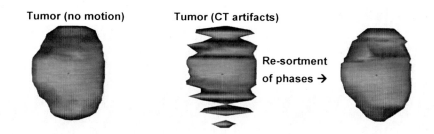

Figure 3. Imaging artifact for a moving tumor. The left hand-side shows the image of a tumor without any organ motion. The middle figure displays a reconstruction of the tumor from CT projections where the motion was not accounted for. The re-sorting of the CT projections to the appropriate breathing phases can restore the right image of the tumor as shown in the right part of the figure.

scanners have to rely on an external breathing monitoring signal, which provides the state of organ motion for the reconstruction process of the CT images.[14, 15] The respective re-sorting of CT projections can usually restore the anatomy of the patient without severe motion artifacts.

The problem of intra-fraction organ motion has to be first accounted for also by the IGRT imaging protocol in the treatment room. However, the quality of 4D-cone beam CTs at the linac lack by far the image quality of 4D-diagnostic CTs . Besides the known problem of enhanced scatter for cone beam CTs, 4D-cone beam CTs at the linac are severely restricted by the number of projections that can be acquired for each breathing-phase.[16] This is mostly caused by the maximal rotational speed of the linac gantry of 360° per minute.

4.2.2. *Organ motion monitoring during treatment*

While the detection of organ motion via x-ray images prior to treatment is already a challenging procedure, the reliable monitoring of organ motion during treatment is even more complicated. Specifically the detection of irregular breathing patterns or unpredictable movements of internal organs like the prostate are difficult.

However, there are several, so called surrogate technologies that do not rely on explicit anatomical images are available for a more or less reliable detection of organ movements during treatment. For breathing induced organ motion these devices range from breathing belts, measurements of the breath temperature, laser or infra-red based measurements of the patient's body surface.

A special device, currently approved only for clinical treatments of prostate tumors, consists of tumor-implanted electromagnetic beacons, whose position in the tumor can be located every tenth of a second.[17]

Depending on the demonstrated reliability of these organ motion monitoring devices their detected motion signals can be used for a dynamic adaptation of treatment parameters during the irradiation process. The two most commonly discussed approaches called 'Gating' and 'Tracking' will be briefly discussed in the following sections.

4.2.3. *Gating of breathing induced organ motion*

The treatment adaptation for the 'gating' technique consists simply in turning the treatment beam on and off while the radiation target can or cannot be detected within a 'window' of anticipated treatment positions, that were selected during treatment planning. This case is illustrated for breathing induced organ motion in Fig. 4. The beam is turned on only during a certain time window close to the exhale phase of the patient. For all remaining tumor positions the beam is switched off.

This simple idea of treatment adaptation for breathing induced organ motion, however, reveals certain disadvantages. First, one has to assume that the organ motion pattern exhibits a certain regularity; otherwise the whole gating process is becoming very inefficient. Furthermore, one has realized that specifically the organ motion of lung tumors is characterized by a large 'inter-fractional' component, the so called base-line shift.[18] This base-line shift has to be accounted for appropriately before any gating strategy will lead to an improved treatment.

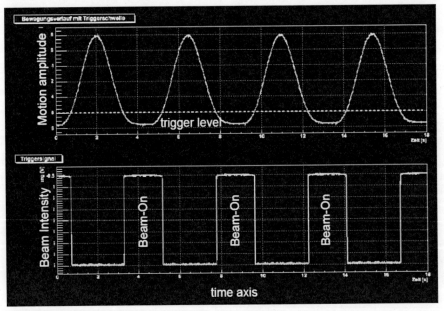

Figure 4. Schematic view of a 'gated' treatment. The signal of a motion monitoring device (upper curve, breathing amplitude) is used to define a trigger level for the Beam-on signal (lower curve).

4.2.4. *Dynamic tumor tracking*

The method of tumor tracking aims to adapt the radiation field in real time to the outline of the moving radiation target. Pre-requisite of this approach is a reliable real-time imaging device that provides the coordinates of the tumor position or even its current velocity. The adaptation of the radiation field for conventional linacs will be accomplished with the multi-leaf collimator. Currently, the algorithms that convert real-time positioning data into adequate velocity profiles of the MLC leaves have been successfully developed, e.g. see ref. McQauid D 2008,[19] Tacke M 2007.[20] The realization of these approaches with standard dose delivery hardware is an on-going topic of research.

4.3. ADAPTIVE INVERSE PLANNING

So far, we have considered adaptive treatment strategies, were it was assumed that the gathered imaging information could be acquired easily and reliably, that the subsequent evaluation and decision process could be performed without any significant time delay and that the resulting adaptation of treatment parameters could also be performed safely on short time scales. However, these assumptions are very demanding. The evaluation and decision procedures have to be based on fast and often automated image processing. These are naturally prone to errors and ambiguities that make an instantaneous decision about the required treatment adaptation difficult, risky or nearly impossible. Moreover, even with the most sophisticated on-line adaption strategy for the treatment, there always remains a residual uncertainty for the considered patient irradiation, which has to be accounted by off-line strategies. The inclusion of the respective sources of uncertainties is usually accomplished in the treatment planning process.

Knowledge about potential geometric treatment uncertainties can be introduced either via safety margin concepts or can be directly integrated in the inverse planning process. Both methods will be briefly discussed in the following.

4.3.1. *Inclusion of uncertainties via margins*

The standard approach to include geometrical uncertainties, like patient setup uncertainties or organ movements, relies on margin concepts as specified within ICRU50[21] and ICRU62.[22] The main focus here is the dose coverage of the clinical target volume (CTV) which should be accomplished by the geometrical concept of the planning target volume (PTV). The design of the PTV provides a sufficient safety margin such that the known uncertainties cannot lead to an under dosage of the CTV for the vast majority of patients. Ideally, the CTV to PTV margin is based on imaging data that allow the derivation of probability distributions for the occurrence geometrical errors. If for instance the variances of the systematic and random target positioning errors are known, standard margin recipes can be used to design the adequate margin for the treatment envisioned.[23]

4.3.2. *Adaptive inverse treatment planning*

However, the derived probabilities for finding a certain anatomical geometry in the treatment reference frame of the linac can be even more efficiently used in the inverse planning process. The idea, for instance, for ensuring a sufficient dose coverage of the CTV is simple. First, if all geometrical uncertainties can be accounted for by probability distributions no PTV margin is assigned to the CTV. A superior dose distribution can now be achieved by delivering more dose to the points in space, where the tumor can be found with a high probability while locations less likely to see the tumor receive less dose. This concept naturally leads to a better dose sparing for adjacent organs at risk.

The workflow for the integration of geometrical uncertainties into the inverse planning process is shown in Fig. 5.

The main feature within the IMRT optimization is that one does not optimize the expectation value of the dose but instead aims for an optimal expectation value of the internal objective function that specifies the quality of a given dose distribution.[24]

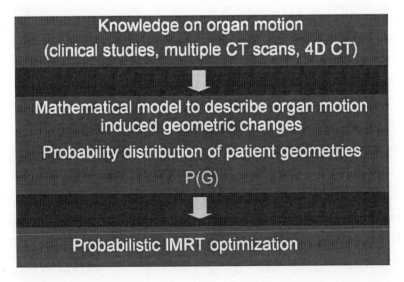

Figure 5. Dataflow for the integration of geometrical uncertainties into the inverse treatment planning process.

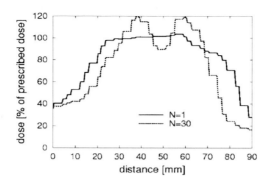

Figure 6. Dose profiles optimized for a various number of fractions in the presence of uncertainties. Details are explained in the main text.

As an example how potential risks influence the results of an adaptive inverse treatment plan, we show dose profiles for the treatment of a prostate tumor in Fig. 6. The optimization was based on a probability distribution for an inter-fraction dislocation of the prostate. Two treatment plans where designed: one for a dose delivery in one fraction and one for a dose delivery in 30 fractions. Figure 6 shows respective dose profiles through the patient where a part of the rectum is encompassed by the prostate. If the dose is delivered in 1 fraction (profile N = 1) then a safe target coverage can only be accomplished by an almost homogeneous dose coverage of the prostate and a significant amount of peripheral healthy tissue also has to be irradiated. For a treatment, however, being delivered in 30 fractions, potential dose errors can easily averaged out and therefore it is possible to spare the central rectum from dose and reducing the amount of irradiated healthy tissue.

References

1. Jaffray DA. Emergent technologies for 3-dimensional image-guided radiation delivery. Semin Radiat Oncol. 2005 Jul; 15(3):208–16. Review.
2. Yan D, Image guided/Adaptive radiotherapy, In: New Technologies in Radiation Oncology, Eds: Schlegel, Bortfeld, Grosu, Springer Berlin Heidelberg New York, 2006.
3. de Boer HC, van Os MJ, Jansen PP, Heijmen BJ. Application of the No Action Level (NAL) protocol to correct for prostate motion based on electronic portal imaging of implanted markers. Int J Radiat Oncol Biol Phys. 2005 Mar 15; 61(4):969–83.
4. Bortfeld T, van Herk M, Jiang SB. When should systematic patient positioning errors in radiotherapy be corrected? Phys Med Biol. 2002 Dec 7; 47(23):N297–302.

5.   Ma CM, Paskalev K. In-room CT techniques for image-guided radiation therapy. Med Dosim. 2006 Spring; 31(1):30–9. Review.

6.   Wong JR, Grimm L, Uematsu M, Oren R, Cheng CW, Merrick S, Schiff P. Image-guided radiotherapy for prostate cancer by CT-linear accelerator combination: prostate movements and dosimetric considerations. Int J Radiat Oncol Biol Phys. 2005 Feb 1; 61(2):561–9.

7.   Jaffray DA, Siewerdsen JH, Wong JW, Martinez AA. Flat-panel cone-beam computed tomography for image-guided radiation therapy. Int J Radiat Oncol Biol Phys. 2002 Aug 1; 53(5):1337–49.

8.   Oelfke U, Tücking T, Nill S, Seeber A, Hesse B, Huber P, Thilmann C. Linac-integrated kV-cone beam CT: technical features and first applications. Med Dosim. 2006 Spring; 31(1):62–70. Review.

9.   Stützel J, Oelfke U, Nill S. A quantitative image quality comparison of four different image guided radiotherapy devices. Radiother Oncol. 2008 Jan; 86(1):20–4.

10. Pouliot J. Megavoltage imaging, megavoltage cone beam CT and dose-guided radiation therapy. Front Radiat Ther Oncol. 2007; 40:132–42. Review.

11. Mackie TR. History of tomotherapy. Phys Med Biol. 2006 Jul 7; 51(13):R427–53. Epub 2006 Jun 20. Review.

12. Sahgal A, Chou D, Ames C, Ma L, Lamborn K, Huang K, Chuang C, Aiken A, Petti P, Weinstein P, Larson D. Image-guided robotic stereotactic body radiotherapy for benign spinal tumors: the University of California San Francisco preliminary experience. Technol Cancer Res Treat. 2007 Dec; 6(6):595–604.

13. Thilmann C, Nill S, Tücking T, Höss A, Hesse B, Dietrich L, Bendl R, Rhein B, Häring P, Thieke C, Oelfke U, Debus J, Huber P. Correction of patient positioning errors based on in-line cone beam CTs: clinical implementation and first experiences. Radiat Oncol. 2006 May 24; 1:16.

14. Vedam SS, Keall PJ, Kini VR, Mostafavi H, Shukla HP, Mohan R. Acquiring a four-dimensional computed tomography dataset using an external respiratory signal. Phys Med Biol. 2003 Jan 7; 48(1):45–62.

15. Keall P. 4-dimensional computed tomography imaging and treatment planning. Semin Radiat Oncol. 2004 Jan; 14(1):81–90. Review.

16. Dietrich L, Jetter S, Tücking T, Nill S, Oelfke U. Linac-integrated 4D cone beam CT: first experimental results. Phys Med Biol. 2006 Jun 7; 51(11):2939–52.

17. Willoughby TR, Kupelian PA, Pouliot J, Shinohara K, Aubin M, Roach M 3rd, Skrumeda LL, Balter JM, Litzenberg DW, Hadley SW, Wei JT, Sandler HM. Target localization and real-time tracking using the Calypso 4D localization system in patients with localized prostate cancer. Int J Radiat Oncol Biol Phys. 2006 Jun 1; 65(2):528–34.

18. Korreman SS, Juhler-Nøttrup T, Boyer AL. Respiratory gated beam delivery cannot facilitate margin reduction, unless combined with respiratory correlated image guidance. Radiother Oncol. 2008 Jan; 86(1):61–8.

19. McQuaid D, Webb S. Target-tracking deliveries using conventional multileaf collimators planned with 4D direct-aperture optimization. Phys Med Biol. 2008 Aug 7; 53(15):4013–29.

20. Tacke M, Nill S, Oelfke U. Real-time tracking of tumor motions and deformations along the leaf travel direction with the aid of a synchronized dynamic MLC leaf sequencer. Phys Med Biol. 2007 Nov 21; 52(22):N505–12.

21. ICRU Report 50, Prescribing, Recording, and Reporting Photon Beam Therapy (1993).

22. ICRU Report 62, Prescribing, Recording and Reporting Photon Beam Therapy (Supplement to ICRU Report 50) (1999).

23. van Herk M, Remeijer P, Rasch C, Lebesque JV. The probability of correct target dosage: dose-population histograms for deriving treatment margins in radiotherapy. Int J Radiat Oncol Biol Phys. 2000 Jul 1; 47(4):1121–35.

24. Unkelbach J, Oelfke U. Inclusion of organ movements in IMRT treatment planning via inverse planning based on probability distributions. Phys Med Biol. 2004 Sep 7; 49(17):4005–29.

25. Unkelbach J, Oelfke U. Incorporating organ movements in IMRT treatment planning for prostate cancer minimizing uncertainties in the inverse planning process. Med Phys. 2005 Aug; 32(8):2471–83.

CLINICAL USE OF 3-D IMAGE REGISTRATION

PETER REMEIJER
The Netherlands Cancer Institute/Antoni van Leeuwenhoek
Hospital, Plesmanlaan 121, 1066CX, Amsterdam,
The Netherlands
Prem@nki.nl

Abstract Using DICOM it has become much easier to exchange images between different manufacturer's equipment, which is essential to clinically implement image registration. Therefore, first a simple introduction to DICOM will be given, in which the advantages of DICOM and some of its deficiencies are explained. Image registration algorithms can be grouped in two main categories: volume registration and chamfer matching algorithms. The merits and drawbacks of both possibilities will be discussed.

Keywords: DICOM; image registration; CT; MRI; PET; cost functions; grey value registration; chamfer matching

1. Introduction

Imaging is a crucial part of any radiotherapy process. Patient dose is planned on CT scans, very often other modalities are included in the tumor delineation process, and recently, image guided procedures involve imaging equipment integrated with the linear accelerator. Two aspects are very important in all this: image transfer and correlating the many different images.

2. DICOM

The basis of most current imaging networks is DICOM, which stands for 'Digital imaging and Communications in Medicine'. It's a very complex standard that addresses all aspects of image storage and communication. The DICOM standard encompasses:

Y. Lemoigne and A. Caner (eds.), *Radiotherapy and Brachytherapy,*
© Springer Science + Business Media B.V. 2009

- A network protocol for communicating medical data
- A logical format for images from any medical source (i.e., CT, MRI) and a physical format on how these images should be stored on disk
- A database structure for medical images

All these aspects are covered in detail in the official DICOM standard.[1] In the following chapters they will be discussed in a simplified way.

2.1. THE DICOM NETWORK PROTOCOL

The DICOM network protocol is very similar to other well known file transfer protocols, like for example 'FTP'. It defines servers and clients, a way to get file listings from the server (like 'dir' in FTP) and move data from the server to a client and vice-versa (like 'get', and 'put' in FTP).

To communicate with a DICOM server or client one identity needs to be known: the AE title. The AE title consists of a combination of the name of the server/client and the port number. The latter is the TCP port used for the DICOM communication. Most applications tend to use port 104, but in principle any port can be used. Table 1 shows an overview of the available file transfer commands within DICOM.

TABLE 1. Possible commands within the DICOM network protocol, compared with FTP commands.

DICOM	FTP	Purpose
C-MOVE	–	Move data from one server to another server
C-FIND	DIR	Find list of files on server
C-STORE	PUT	Send data to server
–	GET	Get data to my PC

One main difference is that DICOM does not know a 'get' command. Therefore, if an image needs to be transferred to a workstation, it will need both 'server' and 'client' functionality. In this way we can then C_MOVE the image from a remote server to our workstation 'server'.

2.2. THE DICOM FILE FORMAT

DICOM stores and transmits most image data in a 2D format (i.e., slice by slice for a CT or MRI scan). An exception is some nuclear medicine data, where a single object can contain 3D or 4D (time resolved) data. The components of each image are rigidly defined by the DICOM standard. An image consists of a large numbers of items that are defined by a group number, an element number, and the data contents. The interpretation and type of the data is defined in a data dictionary that defines all possible group/element number pairs (there are thousands defined, but many are optional or vendor specific). These data contains (amongst others): generic image information, patient information, acquisition information, orientation information, image information, and pixel data.

2.3. THE DICOM DATABASE FORMAT

DICOM also specifies a hierarchical database organization for an image archive and that it should contain the following elements: patients, studies, series and images. A study usually contains all scans made on a particular patient visit on a given scanner. A series usually contains all slices from one 3D volume. Note that the interpretation of a series and a study may differ between manufacturers and modalities. For example, proton density and T2 image pairs acquired simultaneously are often combined in a single series.

2.4. DICOM-RT

A number of DICOM objects have been specifically defined for radiotherapy, and are known as DICOM-RT objects. They should be seen as a subset of DICOM; if a DICOM object is a mammal, then the DICOM-RT object would be a lion (or a cat or…).

Presently, four DICOM_RT objects are distinguished: RT_STRUCT, which are delineated contours or points of interest, RT_IMAGE, which are DRRs or simulator images, RT_PLAN, which is a DICOM representation of a plan and finally, RT_DOSE, which is a 3D grid containing dose values.

In principle, any DICOM storage system (i.e., a PACS) should be able to store DICOM-RT objects. However, almost no system can display these objects because they are usually geared towards radiology use. For that purpose one will still need dedicated RT viewer applications.

3. Image registration and fusion

Image registration is used widely for treatment planning, organ motion studies, image guidance, and follow-up. The purpose of image registration is to find the transformation (translation, rotation, deformation) that maps one scan onto another. In this way, scans can be combined and shown on a pixel-by-pixel basis (e.g., for target volume delineation), or differences can be quantified (for image guidance and follow-up). In radiotherapy, image registration is mostly used to align rigid structures, e.g., bone, in multiple scans. Bone acts as a frame of reference for treatment (verified by means of x-ray images) relative to which the position of organs of interest can be determined. For deformable organs, an alternative could be to use elastic registration. In such a case one scan would be considered as a golden standard to which to the other scan would be warped.

Image fusion is the process of combining the two scans in a single view. For example, by showing a CT scan in grey values, and a PET scan as a color overlay. Hence, image registration and image fusion are different concepts. However, the two terms that are often interchanged.

In the next sections image registration and fusion will be discussed, and a number of other requirements for image registration software will be discussed.

3.1. CHAMFER MATCHING

Chamfer matching is a registration procedure that focuses on bony anatomy or other segmented features.[2-4] This is accomplished by first segmenting features from the images that need to be registered (usually bony anatomy) and then to minimize the distance between the segmented features through the use of a distance map, or distance transform. A distance transform is a 2D or 3D image where each pixel contains the nearest distance to the segmented features, e.g., below the segmented features, all pixel values will be zero (Fig. 1).

The distance between segmented features from the two images is then established by moving the first feature set over the distance map of the second feature set and summing all pixel values below it. Using a cost function minimization algorithm (e.g., the Simplex method) one can then find the optimal transform between the two images by minimizing this distance.

3.2. GREY VALUE REGISTRATION

Grey value registration does not look at distances between segmented features but at grey values on a pixel-by-pixel basis.[5-9] The grey value differences are evaluated through a cost function, and minimization of this cost function by optimizing the relative position of the two images will yield the transform.

Some well known cost functions are: mutual information, correlation ratio, and RMS difference. RMS is the simplest cost function of the three; all pixel differences are squared and summed, and subsequently divided by the number of pixels. Finally, the square root of the sum is taken (Fig. 2). A major drawback of this method is that it only works for equal modalities. Therefore, it is not implemented very often.

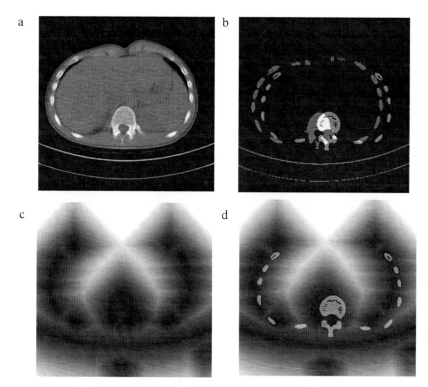

Figure 1. The principle of a distance transform. First, features are segmented from the scan (a), yielding two contour sets (b). From one of the contour sets (the purple one) a distance transform is constructed (c). Finally pixel values under the second contour are summed to determine the distance between the contours.

The correlation ratio cost function looks at correlations between the *pixel values* in both scans, while the mutual information cost function evaluates the correlation between the *histograms* of the images. Both methods are not dependent on the modality and will be able to register e.g., CT to MRI.

3.3. FUSION

Fusion is the process of displaying two (or more) registered images to the user. It is an important step because (a) assessing whether or not we have a 'good' match will depend on it, and (b) the end user (usually a doctor) will need optimal viewing tools. Good evaluation tools are 'sliding windows' and 'color overlay' tools in any orientation. Sliding windows are very good for detecting small errors on a local level, while overlays will be less accurate, but give you a quick and complete overview of the whole image. A few fusion examples are given in Fig. 3.

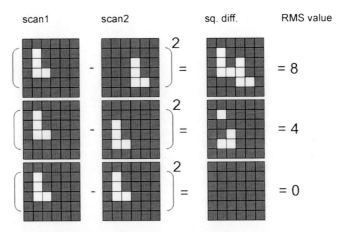

Figure 2. Grey value registration, using the RMS cost function as an example. The large squares represent two bi-colored images, with the small squares representing the pixels (grey = 0, white = 1). The three rows represent three possible relative positions of the two scans. The top row is not aligned at all, the middle row is partly aligned, and the lower row is fully aligned.

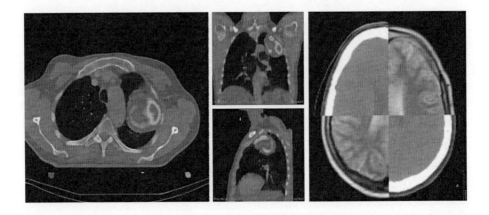

Figure 3. Examples of image fusion. The left image shows PET-CT fusion. The PET signal is shown as a color overlay on top of the CT scan. This works well because PET is a low resolution image (only used as an indicator to know if e.g., a lymph node is positive) and geographical information is taken from the CT image. The right image shows 'sliding window' fusion of a CT and MR scan. See color picture in Appendix I.

3.4. ADDITIONAL REGISTRATION TOOLS

There are a number of tools that are invaluable for clinical use of 3D image registration and should be part of any image registration software. First is the ability to do a rough pre-match, either manually or by, for example, aligning the centers of gravity of the two images. The reason for this is that most registration algorithms can only align two images that are misaligned within a certain range; usually they should be within 1 cm from each other.

Another important tool is the possibility to indicate regions of interest in the images that are used in the algorithm. In this way, it is possible to exclude mobile anatomy from the registration process (e.g., a moving leg), which improves the accuracy.

Finally, although DICOM can be used to store the images in a PACS, this is not the case for the registration results. For these data, a separate database will be needed, and precautions regarding backups and archiving need to be taken to make sure these data will be available later on.

4. Conclusion

Registration and fusion are important aspects in modern radiotherapy and invaluable for accurate delineation for a number of treatment sites and image guided radiotherapy. DICOM has simplified image handling to a large extent, but separate databases are still needed to store registration results.

References

1. http//dicom.nema.org
2. van Herk M, Kooy HM (1994) Automatic three-dimensional correlation of CT-CT, CT-MRI, and CT-SPECT using chamfer matching. Med Phys 21:1163–78.
3. Kooy HM, van Herk M, Barnes PD, Alexander E 3rd, Dunbar SF, Tarbell NJ, Mulkern RV, Holupka EJ, Loeffler JS (1994) Image fusion for stereotactic radiotherapy and radiosurgery treatment planning. Int J Radiat Oncol Biol Phys 28:1229–34.
4. Mangin JF, Frouin V, Bloch I, Bendriem B, Lopez-Krahe J (1994) Fast nonsupervised 3D registration of PET and MR images of the brain. J Cereb Blood Flow Metab 14:749–62.
5. Maes F, Collignon A, Vandermeulen D, Marchal G, Suetens P (1997) Multimodality image registration by maximization of mutual information. IEEE Trans Med Imaging 16(2):187–98.
6. Grimson WL, Ettinger GJ, White SJ, Lozano-Perez T, Wells WM, Kikinis R (1996) An automatic registration method for frameless stereotaxy, image guided surgery, and enhanced reality visualization. IEEE Trans Med Imaging 15:129–40.
7. Hajnal JV, Saeed N, Soar EJ, Oatridge A, Young IR, Bydder GM (1995) A registration and interpolation procedure for subvoxel matching of serially acquired MR images. J Comput Assist Tomogr 19:289–96.
8. Hill DL, Batchelor PG, Holden M, Hawkes DJ (2001) Medical image registration. Phys Med Biol 46:R1–45.
9. Studholme C, Hill DL, Hawkes DJ (1997) Automated three-dimensional registration of magnetic resonance and positron emission tomography brain images by multiresolution optimization of voxel similarity measures. Med Phys 24:25–35.

GEOMETRICAL UNCERTAINTIES IN RADIOTHERAPY

PETER REMEIJER
The Netherlands Cancer Institute – Antoni van Leeuwenhoek Hospital, Plesmanlaan 121, 1066CX, Amsterdam, The Netherlands
Prem@nki.nl

Abstract Geometrical uncertainties are a fact in any radiotherapy practice. Lasers can be misaligned, patients are mobile and the definition of the target volume is not always very easy. To deal with these uncertainties a safety margin is applied, i.e. a larger volume than the target itself is treated. In this chapter we will discuss common sources of geometrical uncertainties and how to compute these safety margins.

Keywords: CTV; PTV; geometry; uncertainty; margin

1. The radiotherapy chain

Figure 1 depicts an overview of the radiotherapy chain in 17 steps.[8] All of these steps will introduce a small error, usually on the order of a millimeter. However, in some cases large errors can occur, when most errors are in the same direction.

1.1. DELINEATION ERRORS

Basically, delineation errors are a misplacement of the delineated contour with respect to the true position of the target volume or tumor. The problem is that it is often very hard to know where that true position is. In many cases, the use of multiple modalities will improve accuracy,[1] but the real 'gold truth' can only be found through pathology. Because delineation is performed only once per patient, any errors introduced in this stage will be systematic. Multi-observer studies can be used to quantify the delineation error [4–6] (Fig. 1).

Y. Lemoigne and A. Caner (eds.), *Radiotherapy and Brachytherapy*,
© Springer Science + Business Media B.V. 2009

P. REMEIJER

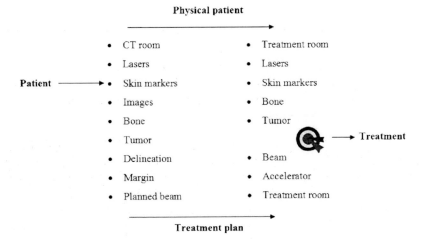

Figure 1. The radiotherapy chain.

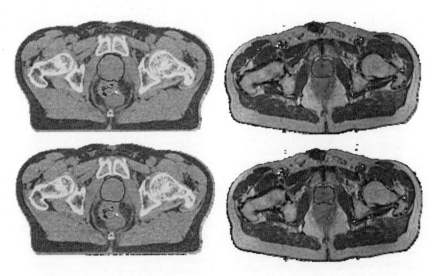

Figure 2. An example of delineation uncertainty. On the left, delineations of the prostate on a CT scan by two observers are shown. The second observer clearly delineated a smaller volume. On the right the same two observers performed the same delineation on MR, revealing an even larger modality (MR versus CT) difference.

1.2. ORGAN MOTION

Organ motion is most commonly defined as motion of the target volume, or tumor, relative to the bony anatomy (Fig. 3). Organ motion is always present and therefore happens both during the scanning and the treatment phase. An example of a systematic organ motion error could be a shifted prostate due to rectum filling at the time of scanning. To quantify organ motion, a repeat study on a group of patients can be performed.

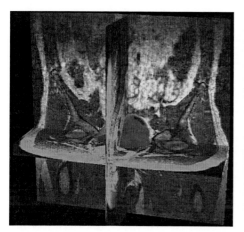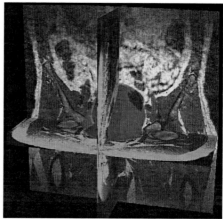

Figure 3. Organ motion. Depicted are two 3-D views of the bladder, taken 1 h apart. The total movement of the cranial bladder wall was 7 cm, which shows that in some cases organ motion will be very difficult to determine a priori. (Color photo in the color section at the end of the book.)

1.3. SETUP ERRORS

Setup errors are displacements of the patient's bony anatomy with respect to the treatment room coordinate system, usually indicated by lasers. These errors also occur both during the treatment preparation phase (i.e. scanning) and during treatment delivery. By monitoring setup errors by, for example, portal imaging it is possible to quantify them.[2, 7]

2. Analysis of errors

Having a closer look at the radiotherapy chain (Fig. 1) we see that certain errors in the treatment will be the same for all patients (e.g. laser misalignment), or the same for a single patient treatment (e.g. delineation error), or be limited to one fraction. We will call these errors group mean (μ), systematic (Σ) and random errors (σ), respectively.

The impact of the three types of errors on the size of the margin is different. Therefore it is imperative to quantify all, in order to be able to compute an 'evidence based' margin.

The starting point for any analysis is to measure deviations for individual fractions. For example, one could measure the setup error in the cranial-caudal, anterior-posterior, and left right direction for each fraction and for a number of patients using portal imaging.

The group mean error is then computed by averaging all measurements of all patients. Within each patient we can compute the mean and the standard deviation of all fractions. This will yield a *patient specific* systematic error and random standard deviation (SD), respectively.

The group systematic error (Σ) is then determined by taking the standard deviation of all patient specific systematic errors. Finally, the group's random error (σ) is computed by taking the RMS value of all patient specific random standard deviations.

3. Margins

The underlying idea of a geometric margin is that *group statistics* are used to predict what errors one can encounter in patients. Subsequently, the margin is chosen in such a way that most errors are covered. This methodology has the advantage that a margin can be determined without the need to measure patient specific errors.

Earlier, it was indicated that different types or error will require different margins. The total margin will then be the sum of the individual margins.

3.1. MARGINS FOR RANDOM ERRORS

The effect of random errors is that the dose will be delivered at different locations in the patient for each fraction, while the *average* position of the dose distribution will be correct. The effect of this will be that the dose distribution gets blurred with respect to the original dose distribution, i.e., the penumbra will become wider.

For many fractions this is mathematically equivalent to a *convolution* of the dose distribution with the random error distribution.[3] The margin that is needed to compensate for the 'blurring' effect is the distance between the 95% dose level of the original distribution and the 95% dose level of the convolved (blurred) dose distribution. For a simplified case of a single beam the margin can be approximated by 0.7σ, where σ is the root-sum-square of the $\sigma - s$ determined for organ motion and set-up. Note that delineation is not included here because it is a purely systematic error.

3.2. MARGINS FOR SYSTEMATIC ERRORS

The systematic error of a particular patient is not known a priori. What *is* known is that the systematic errors are from a distribution of errors with standard deviation Σ. This means that a confidence interval can be defined with a certain confidence level, e.g., 90% of all errors will be within $-2\Sigma....2\Sigma$.

To define a confidence interval, two things are needed. First, the confidence level needs to be chosen, and second, the probability distribution needs to be known.

For the confidence level we will choose a, somewhat arbitrary, level of 90%, i.e. we want a margin that covers 90% of all patients. Because we deal with *3-D* systematic errors in radiotherapy, a 3-D probability distribution, known as the chi-squared (χ^2) distribution is needed. The margin for systematic errors is then determined by the 90% confidence interval for this χ^2 distribution: 2.5Σ, where Σ is the root-sum-square of all Σ – s determined for delineation, organ motion and set-up.

3.3. COMBINING IT ALL

The computation of the total margin is now very straightforward; just add the individual margins that were determined for the random and systematic errors:

$$\text{Margin} = 2.5\Sigma + 0.7\sigma$$

Note that the overall mean (μ) does not appear in the margin formula. The reason for this is that this is an error which is *the same for all patients*. Hence it is possible to correct for this error in advance. Furthermore, this formula assumes a rigid body of reasonably large size, i.e. larger than the 'penumbra size'.

3.4. HOW TO REDUCE THE MARGINS?

The multiplication factor preceding Σ is about a factor four larger than the factor preceding σ. This implies that systematic errors have a fourfold larger effect on the margin than random errors.

Therefore, the best way to reduce margins is to focus on systematic errors first. There are a number of ways to do this. Portal imaging can be used to determine setup errors and subsequently correct by an average displacement. By using multiple modalities, delineation uncertainties can be reduced and hence the systematic errors will be too. Changes in the position of the tumor, relative to its position in the planning CT scan can be monitored by, for example, repeat CT scans in the first week of treatment. Any changes could then be incorporated in an adapted plan that is deployed from the second week onward. This procedure is known as 'adaptive radiotherapy'.[7]

To reduce the margin even further, the random errors can be corrected as well. However, this requires daily measurements and online corrections, which is not always feasible from a workload or workflow point of view.

4. Conclusion

Geometrical uncertainties in radiotherapy are unavoidable and should be dealt with by using a safety margin. The only way to determine this margin is by quantifying the statistics of all errors involved.

A reduction of the margin can be achieved most effectively by reducing the systematic errors. Because this usually only involves measuring the errors over a couple of fractions and offline analysis, the amount of additional linac time required for such a procedure is very limited. Organ motion will mostly be the predominant cause for geometrical uncertainties, but do not underestimate the effect of delineation uncertainties, especially because these are purely systematic errors.

References

1. Aoyama H, Shirato H, Nishioka T, Hashimoto S, Tsuchiya K, Kagei K, Onimaru R, Watanabe Y, Miyasaka K. Magnetic resonance imaging system for three-dimensional conformal radiotherapy and its impact on gross tumor volume delineation of central nervous system tumors. Int J Radiat Oncol Biol Phys 2001 Jul 1; 50(3):821–827.
2. Bel A, van Herk, M, Lebesque JV. Target margins for random geometrical treatment uncertainties in conformal radiotherapy. Med Phys 1996; 23:1537–1545.
3. Bijhold J, Lebesque JV, Hart AAM, Vijlbrief RE. Maximizing setup accuracy using portal images as applied to a conformal boost technique for prostatic cancer. Radiother Oncol 1992; 24:261–271.
4. Court LE, Dong L, Taylor N, Ballo M, Kitamura K, Lee AK, O'Daniel J, White RA, Cheung R, Kuban D. Evaluation of a contour-alignment technique for CT-guided prostate radiotherapy: an intra- and interobserver study. Int J Radiat Oncol Biol Phys 2004 Jun 1; 59(2):412–418.
5. Rasch C, Barillot I, Remeijer P, et al. Definition of the prostate in CT and MRI: a multiobserver study. Int J Radiat Oncol Biol Phys 1999; 43:57–66.
6. Remeijer P, Rasch, C, Lebesque, JV, et al. A general methodology for three-dimensional analysis of variation in target volume delineation. Med Phys 1999; 26:931–940.
7. van Herk M, Remeijer P, Rasch C, Lebesque JV. The probability of correct target dosage - dose population histograms for deriving treatment margins in radiotherapy. Int J Radiat Oncol Biol Phys 2000 Jul 1; 47(4):1121–1135.
8. Yan D, Wong JW, Vicini F, Michalski J, Pan C, Frazier A, Horwitz E, Martinez A. Adaptive modification of treatment planning to minimize the deleterious effects of treatment setup errors. Int J Radiat Oncol Biol Phys 1997; 38:197–206.

DELIVERY OF INTENSITY-MODULATED RADIATION THERAPY INCLUDING COMPENSATION FOR ORGAN MOTION

STEVE WEBB
Joint Department of Physics and Royal Marsden NHS Foundation Trust, Downs Road, Sutton, Surrey, SM2 5PT, London, UK
steve.webb@icr.ac.uk.

Abstract In this chapter, covering 3 h of lectures at ESMP, the topics reviewed are:

- The design and use of the multileaf collimator (MLC)
- The requirement and rationale for intensity-modulated radiation therapy (IMRT)
- The techniques to deliver IMRT with a MLC
- The techniques to deliver IMRT without a MLC
- Techniques to compensate for organ tissue motion

Keywords: MLC, IMRT; organ tissue motion

1. Preamble

The first ESMP was held in Archamps in November 1998 just at a time when the methods to plan and deliver IMRT were moving from the research-hospital "one-off" stage into a commercially-supported product and IMRT was therefore only being performed at a very limited number of Centres. At the time of writing, between the 10th and the 11th ESMP, the IMRT delivery techniques are mature clinical products, supported by commercial industrial equipment and are being introduced in both clinical trials and routinely. This topic has been taught at all ten ESMP Schools year-by-year and the material has been changed periodically to reflect current practice. Whilst it is not expected that the basic methods will evolve greatly for treating an assumed-static patient, they will indeed evolve for treating the moving patient and so the material presented should be supplemented by reading the latest research papers. Most of the lecture material can already be found in four books[7-10] and the figures here are adapted from those in the books and as circulated to students each year at

ESMP. This chapter is a brief resume of this material. Other excellent reference works in book or DVD form are Van Dyk,[6] Palta and Mackie,[4] Mundt and Roeske (2005), Bortfeld et al.,[1] Mayles et al.[2] and Schlegel and Mahr.[5]

2. The design and use of the multileaf collimator (MLC)

The goal of radiation therapy is articulated differently by people from different backgrounds. A patient would say "I want to get better from my cancer with minimal side effects". The medical doctor would translate this to "Try to maximise the tumour-control probability minimising normal-tissue complication probability" and the physicist would further translate to "Make the dose as high (and maybe as uniform) as possible in the tumour target and as low as possible in the organs at risk". In two verbal "moves" the goal has gone from one of human need to a technical requirement in terms of distributing a quantity (dose) of clear physical units and dimensions.

This goal is not new; in fact it was articulated in 1896 when, just a year after the discovery of the x-ray, radiotherapy started. The difference between now and then is that today we have ways to solve the problems. Let us pass over 50 years of radiotherapy in which largely a few beams from a few fixed directions were made to overlap in the patient and create a dose that was "concentrated" at the target. Such distributions were box-like (for fixed-field therapy or cylinder-like (for rotation therapy) and all these primitive distributions of dose over-irradiated normal tissue and did not optimally treat the tumour target. The techniques which have been developed in the last 20 years effectively aim to minimise dose to unwanted normal tissues by creating a dose volume that tightly conforms to the target volume. We can think of this as if the high-dose volume were a "cling film" of high dose which is shrink wrapped onto the target. Techniques to do this began in the late 1980s and are generically known as conformal therapy. Intensity-modulated radiation therapy (IMRT) is a subset and finessing of the technique of conformal therapy.

The first priority of conformal therapy is to geometrically shape the radiation field to the so-called beam's-eye view of the target and in this way to keep the organs at risk out of the field of view of the primary beam.

We may recall that such a goal was addressed as early as the mid 1960s by techniques of gravity-oriented blocking and dynamic field shaping. In these techniques, fields were rotated relative to the recumbent static patient and the aperture of the field was dynamically adjusted by cams such that the beam's-eye view of the target was reflected in the shape of the field generated by the dynamic cams. At the same time, rotating absorbers made of heavy metal and of minified shape, with respect to the organs at risk, were also rotated in order to maintain the organs at risk always in the geometric shadow of such absorbers. These techniques led to some of the first isodose distributions which had concavities in their surface.

A primitive form of multileaf collimator (MLC) was patented as early as 1959. This MLC comprised four banks of "leaves" which in practice were large blocks of attenuating tungsten. These blocks were moved into and out of the field using mechanical knobs attached to rods which drove camshafts which pushed the rods into the field and pulled them back out again. In this way very primitive stepped-edged field shapes could be created. Such a primitive multileaf collimator was, it is believed, actually constructed by Toshiba. The concept of the modern multileaf collimator was first articulated in a prototype built by Scanditronics in about 1984. This comprised of series of tungsten leaves, controlled by motors, which in turn were controlled by a personal computer. The leaves could be moved in and out of the radiation area to create a field which was step shaped. The initial concept was that each leaf pair would create a strip of radiation that corresponded to the projection of a target-volume strip as extracted from a CT slice. The first such presentation was seen by the author in Jerusalem at the ESTRO-3 Conference in 1984. Throughout the ESMP teaching, the author has made use of a plastic multileaf collimator for real-time demonstrations and showed slides of much more complex MLCs that are in use today.

Historically the first multileaf collimators were for collimating the radiation from diagnostic x-ray tubes and patents showed diagrams of MLCs as long ago as 1906. A patent of 1959 appears to have been largely overlooked and modern developments can be dated from the mid 1980s.

The multileaf collimator, whilst superficially a simple device, can be characterised by a number of individual properties. These include: (1) the number of leaf pairs, (2) the width of the leaves at isocentre, (3) the compatability with accelerator type, (4) maximum field size and (5) "over travel" distance. The reader is referred to more detailed textbooks for tables which show the parameters for individual specific MLCs and, as the years pass by, more and more such devices are put on the market with new and supposedly compelling properties.

The 1985 patent for the modern MLC is held by Brahme. This patent showed leaves that are double-focused, driven by independent motors, provided with stepped ends and sides and monitored by a television arrangement. Double focusing is the name given to the geometry whereby the leaves move on an arc of a circle both in their along-leaf and in their across-leaf directions so that the leaf penumbra of the field is effectively constant with field size. The stepped ends are to avoid radiation leakage and the sides of the leaves are also stepped for the same reason. The major commercial manufactures of MLCs are Varian, Elekta, Siemens, Scanditronics, Mitsubishi and General Electric.

The simplest use of an MLC is to shape the field either statically or dynamically. This means that the field shape is adjusted to the beam's-eye view of the target from each and every field direction. For this reason the leaves need to be under fine computer control.

At ESMP slides were shown of a variety of commercial and home-made multileaf collimators. Notably amongst the latter is the micro MLC designed and constructed at DKFZ, Heidelberg. This can be manually attached to the head of a linear accelerator in the same way as a gamma-camera collimator is attached to the head of a gamma camera and simultaneously the field shape can be adjusted by manually setting the leaves to a plug of wood or Perspex that is subsequently removed prior to irradiation. The group at Heidelberg are famous for designing a large number of MLCs culminating in one with some very specific properties whereby the leaf end can vary its orientation with respect to the leaf as the leaves move across the field. This is a design feature meant to minimise penumbra changes with field-shape change. The way this is done is to make the leaf and the leaf end be controlled by a single cog that rotates over a pair of toothed racks, each rack having a slightly different number of teeth per inch. It is an ingenious design.

The question arises of how a multileaf collimator should be oriented in relation to the beam's-eye view projection of a target volume. This problem was addressed by Brahme very early on who showed that, in order to minimise the excess region of treated tissue, the leaves should be arranged such that they are at right angles to the largest dimensions of the planning target volume, that is the side of the planning target volume with the greatest curvature. This theory of orientation can be proved mathematically but to some extent is intuitively obvious. There is no problem doing this for the majority of convex target volumes. The problem becomes insoluble for concave target volumes when the concavity becomes such that no single field can shape the projection of the target area. In that case it is necessary to join together several fields and this in itself creates problems because the joining of fields creates what is known is a tongue-and-groove underdose. The tongue-and-groove underdose arises because, at the junction of two fields, only the edge of the leaves would have been in use for shaping the field and when two edged distributions are added together they do not create the same as an open field. However, when electron transport is taken into account it turns out that the underdose is more spread out and less in magnitude than the theoretical magnitude for primary fluence.

The next issue with respect to a multileaf collimator is to decide the options for placing such a collimator with respect to the edges of the projected planning target volume. In general a field edge, defined by a multileaf collimator, will have stepped edges. There are, therefore, three possibilities. Either the internal points of such stepped edges can be aligned with the field edge; this would definitely give full protection of the tissues surrounding the PTV but it will under-irradiate the periphery of the PTV. The opposite alternative is to put the external cusps of the toothed MLC adjacent to the field edge. This will create a planning target volume dose which is not under-dosed but would provide some dose to unwanted organs at risk surrounding the planning target volume. The third option is, of course, the compromise known as "transection" in which the mid points of the toothed edges are aligned along the field edge.

The advantages of an MLC over conventional blocking are many. It leads to cleaner environment, a reduced requirement for storing blocks, the elimination of the need to lift heavy lead blocks, the elimination of the need to enter a room to change fields, and the elimination of toxic casting. The field shapes could be more easily verified and they are less susceptible to human error. Finally, with respect to the multileaf collimator, once an MLC is purchased and commissioned there are a number of detailed studies that have to be performed to characterise the MLC. The studies should include a quantification of penumbra as a function of leaf end orientation. There should be a check that the depth-dose curves and the output factors are equivalent to the use of jaw-collimated fields. Leakage should be measured through the leaves, between the leaves and between the leaf ends. The output factors for small fields should be measured. The leakage when an MLC is combined with the use of back-up diaphragms if used should be checked and the reproducibility of leaf positions should be checked.

It has now generally been accepted that an MLC can replace cast blocks for most situations. There are a few special situations which still require use of lead blocks. Verification of the leaf patterns generally relies on the use of either a TV system or the playback of digital records of the leaf positions and/or the use of portal imaging devices.

3. The requirement and rationale for intensity-modulated radiation therapy (IMRT)

Conformal radiotherapy by geometrical shaping only can only fit a high-dose volume around a 3D shape which has a convex surface. Conformal radiotherapy with intensity-modulated (and geometrically-shaped) fields can give a 3D high-dose volume with concavities in its surface. This is the major step forward for high-precision radiotherapy that can lead to an improvement in the tumour-control probability without damaging the normal-tissue-complication probability. We may imagine the body as a set of slices irradiated by two-dimensional beams that may be considered as a set of one-dimensional beams irradiating each slice (to first order). Each one-dimensional beam is then divided into beam elements (bixels), each with a different intensity. The superposition of such beams leads to a high-dose volume with concave border. Although simply-blocked fields and wedged fields do introduce variability in intensity across the field, they are generally not regarded as true IMRT. We may imagine scenarios in which the intensity is stratified (quantised) across the field at different spatial intervals. These quantisations and spatial intervals could be coarse or fine. If both are fine we have the situation of true IMRT. What is the clinical rational for IMRT? Three examples may be immediately conceptualised. The first is the male pro-state, an organ which varies its cross section with longitudinal position within the body. The rectum impinges posteriorly on the organ and the bladder impinges anteriorly on the organ. The rectum and bladder are organs at risk. Hence irradiating the prostate requires IMRT.

A second example is the irradiation of concave volumes in the brain in which the goal is to spare normal structures such as the orbits, the optic nerves and the brain stem. Finally, we may imagine a class of head-and-neck cancer problems such as the irradiation of thyroid tumours with involved neck nodes in which the challenge is to keep the dose to the spinal cord low and also the dose to the oesophagus low. Such cases are complex planning cases and require IMRT.

When we may consider that two-dimensional intensity-modulated beams (IMBs) may be made as the sum of a set of one-dimensional beams, it becomes apparent that there are at least six methods to generate such 1D IMBs. These are: (1) compensators, (2) multiple exposures with different static MLC settings, (3) dynamic sweeping of MLC leaves, (4) apparatus for Tomotherapy, (5) electronically-steered time-modulated pencil beams, (6) the moving-attenuating-bar technique. The lectures at the ESMP concentrated on methods 2, 3 and 4 since these are the popular clinical techniques.

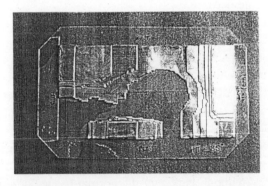

Figure 1. A photograph of a metal compensator created by cutting thin sheets of lead and glueing them together. This was how compensators were made for a trial of breast IMRT at the Royal Marsden Hospital. (Courtesy of Dr. P. Evans.)

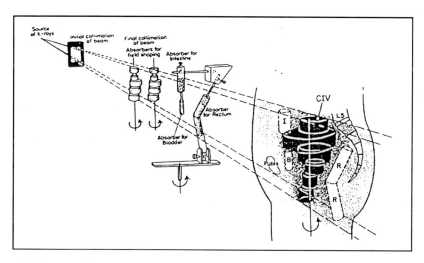

Figure 2. How synchronous absorbers for field shaping and synchronous protection can be arranged to provide a high-dose volume in the continuously-irradiated volume (CIV) and protect the intestine (I), bladder (B), and rectum (R).

4. The techniques to deliver IMRT with a MLC

The simplest technique to deliver a 1D IMB using a MLC is called the multiple-static-field (MSF) technique. The MSF (MLC) IMRT technique was invented by Thomas Bortfeld and Art Boyer during the period when Bortfeld was in Texas in 1993. The idea is that each leaf pair of a MLC defines a one-dimensional IMB. The one-dimensional IMB is then built up as a series of independent static-field components. It is convenient to consider these to be of equal fluence increment. The radiation is off between re-setting leaf positions and on only when the leaves have reached their new positions. With this scenario the set of leaf pairs all function independently.

In the early days, the technique was performed manually and was very inefficient because the total treatment time included all the beam-off times. Today manufacturers (for example Varian, Siemens, Elekta) have automated this process and it gains the name "step and shoot" or "stop and shoot."

If the 1D-modulated field is monotonic (that is has only one peak) and is segmented such that there are N rising steps and N falling steps, it is self evident that there are N! possible ways in which the field components may be made. In practice only two of this N! number of combinations are ever engineered. The first is the leaf-sweep technique, in which the first leading edge is paired with the first trailing edge; the second leading edge is paired with the second trailing edge and so on until the N-th leading edge is paired with the N-th trailing edge. In this scenario the leaves move unidirectionally from one side of the field to another ensuring there is no backlash in any of the driving gear arrangements. The alternative is called leaf close-in in which the leaves are paired in such a way as to close in on the various peaks in the distribution. It could be appreciated that in this scenario the leaves have to move bi-directionally and it is because of the potential for error in doing this that this scenario is not normally engineered.

More detailed research papers have looked at the situation when there is more than one peak in the one-dimensional IMB and detailed mathematical formulae exist to predict the number of combinations. These combinations may be exploited to produce two-dimensional fields that have minimum tongue-and-groove underdose.

The first manual experiments were done by Thomas Bortfeld in Houston who created nine two-dimensional fields around a polystyrene phantom comprising film sandwiched between the slices of the phantom intended to represent a case of prostate IMRT. Between 21 and 30 fluence increments were used and nine slices were engineered. The irradiation was made with a Varian accelerator and MLC and the films so irradiated were then digitised and manual isodose contours were drawn to show the high conformality of dose. This was the

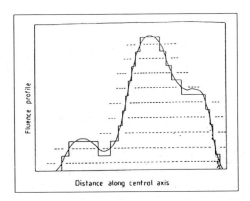

Figure 3. Shows how a 1-D intensity modulation may be created for a radiotherapy beam profile. The horizontal axis is the distance along the direction of travel of the leaf, measured at the isocentre of the beam (called a central axis in the transaxial cross-section of the patient). The vertical axis is x-ray fluence. The solid line is the intensity modulation expressed as a continuous function of distance, interpolated from the discrete modulation resulting from some method of inverse planning. The horizontal dotted lines are the discrete intervals of fluence. Vertical lines are created where the dotted lines intersect the continuous profile thus giving a set of discrete distances at which discrete fluence increments or decrements take place. These are realised by setting the left and right leaves of a MLC leaf-pair at these distances in either "close-in" or "leaf-sweep" technique. Note all left leaf settings occur at positions where the fluence is increasing and all right leaf settings occur at positions where the fluence is decreasing.

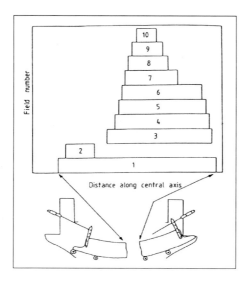

Figure 4. Shows the ten separate fields which when combined would give the distribution of
fluence shown in Fig. 3. Each rectangle represents a field and the left vertical edge is the position
of the left leaf and the right vertical edge is the position of the left leaf and the right vertical edge
is the position of the right leaf. This method of setting the leaves is known as the "close-in"
technique. A schematic of a pair of MLC leaves is shown below the fields with arrows indicating
the correspondence with the field edges.

first time this had been done anywhere in the world and the work won the
prize for the best paper at the 11th ICCR in Manchester in 1994. Today the
engineering of this technique is virtually taken for granted and commercial MLCs
on commercial linear accelerators can deliver two-dimensional fields this way
by multiple-static-field "step and shoot".

Now we turn to the dynamic MLC technique based on the use of a MLC.
The dynamic MLC technique (DMLC) is sometimes called the camera-shutter
technique because old single-lens-reflex cameras had shutters that moved like
the leaves of a MLC. An alternative name is leaf-chasing or sliding window.

Clearly such a technique requires that the leaves be driven by motors, that
there be methods to measure and verify the leaf position and that consideration
is given to anti-collision and overtravel and the tongue-and-groove effect.
There is no unique solution for the leaf velocities which yield a required 1D
IMB even considering only the primary fluence. However, three research teams
independently discovered in 1994 the algorithm which minimises the treatment
time.The key players were Stein et al. at DKFZ, Heidelberg, Spirou and Chui at
Memorial Sloan Kettering Cancer Center, New York; and Svensson et al. at the
Karolinska Institute in Stockholm.

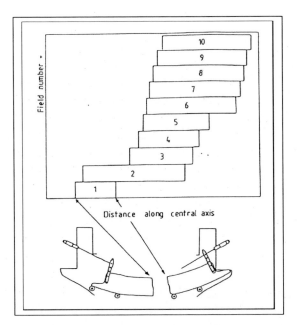

Figure 5. Shows the ten separate fields which when combined would give the distribution of fluence shown in Fig. 3. Each rectangle represents a field and the left vertical edge is the position of the left leaf and the right vertical edge is the position of the left leaf and the right vertical edge is the position of the right leaf. This method of setting the leaves is known as the "leaf-sweep" technique. A schematic of a pair of MLC leaves is shown below the fields with arrows indicating the correspondence with the field edges.

The papers were going through the refereeing process simultaneously and it would be invidious to consider that one of these may be regarded as the unique inventor of the optimum leaf-speed algorithm. They all came up with the same algorithm.

The algorithm required that the leaves need to move slowest where the rate of change of intensity is largest and vice versa, an initially counterintuitive observation. Hence, slowly varying IMBs are more difficult to generate by the DMLC technique than rapidly varying IMBs.

The concept of time-position diagrams was introduced in which, if one focuses at any leaf position, the difference between the trajectory of the leading and the trailing leaf corresponds to the delivered primary intensity. The major discovery of the three groups was that when the gradient of the intensity profile is positive the leading leaf should move at the maximum velocity $\hat{v}$ and the trailing leaf would provide the modulation. Conversely when the gradient of the intensity profile is negative the trailing leaf should move at the maximum velocity $\hat{v}$ and the leading leaf should provide the modulation.

The papers also formally proved that no other algorithm provided optimum treatment time. The paper by Svensson showed what to do if the maximum velocities or accelerations were exceeded and, in general, it was found that, depending on the complexity of the IMB, the efficiency would be less than one and often around ⅓. More complex envisions of the technique allowed for upstream spatial intensity modulation which could decrease the overall treatment time. Stein in particularly showed how to iteratively build in the in-plane scatter of the leaf penumbra using intermediate virtual profiles and through an interactive technique obtained convergence of the delivered beam to the required beam. Spirou and Chui also showed that the leaf leakage could be used to advantage in the DMLC technique by creating pseudo-intensity profiles to which the leaf positions were fitted, such that the transmission fluence was used constructively in the building up of the delivered IMB.

The DMLC technique has widely been used and in particular, despite several manufacturers reverting to the MSF-MLC multiple-static-field technique a further reversion to the DMLC technique would allow for specific methods of compensating for organ motion through the use of "breathing leaves". This topic will be described in section 6.

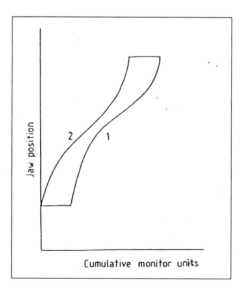

Figure 6. Showing the jaw or leaf positions as a function of cumulative monitor units for two jaws or leaves (the trailing jaw or leaf is number 1) in dynamic therapy. The horizontal axis is a measure of cumulative MUs representing time t. The vertical axis is a measure of position x. At any position x, the horizontal width between the two curves gives the intensity of the IMB in MU.

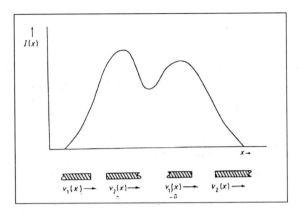

Figure 7. When the gradient dI/dx of the intensity profile I(x) is positive, the leading leaf (2) should move at the maximum velocity; conversely when the gradient dI/dx of the intensity profile I(x) is negative the trailing leaf (1) should move at the maximum velocity. Equation (2) is illustrated by showing a schematic of the pair of leaves in two separate locations delivering the IMB profile shown in the upper part of the figure.

One very old IMRT technique is intensity-modulated arc therapy (IMAT) which was developed by Yu when in the William Beaumont Hospital, Detroit around 1995. In this technique the gantry makes multiple rotations. The theory is that, at each gantry orientation of each rotation, one of the components is picked off from the decomposition of the modulated field at that orientation and this is done in such a way as to minimise the change of MLC leaf position between each orientation. After the full number of rotations are completed the modulation has been successfully built up. The along-the-leaf spatial resolution is continuous; it is an efficient photon use and there are no match-line problems. This technique lay dormant for nearly a decade because there was no commercial treatment-planning system for it but it has recently been subject to a resurgence of interest particularly with Elekta and Varian now offering versions of this technique known as VMAT (volumetric-modulated-arc therapy) and RapidArc respectively. In this technique the variability of the fluence is even greater than with IMAT. It is possible to vary both the fluence output rate and the gantry speed during the treatment in such a way that multiple arcs are not so necessary and efficient IMRT can be delivered with just a single or a pair of arcs which may be co-planar or non-co-planar.

With all MLC-based IMRT techniques there are a number of key issues to address. The leaf setting generally requires some kind of interpreter and a whole interpreter-generating industry operated between the mid 1990s and the early 2000s. Generally also a virtual DMLC is provided, being an emulator that allows the user to view the leaf movement and the building up of fluence. Attention has to be paid to the dose calculation and this requires a model of the dose from variable-sized fields. Finally, verification is important to ensure that the leaves had reached the positions that they were designed to reach and many integrating portal imaging systems can yield integral fluence profiles for such verification purposes. Generally it is necessary to modify the leaf positions in an interpreter to take account of head scatter, transmission and phantom scatter in order that measured three-dimensional dose distributions match the calculated three-dimensional dose distributions.

These techniques have been applied widely to a number of different cancer sites including head and neck cancer, breast cancer and prostate cancer.

Regarding verification, techniques have evolved over the years. The very simplest verification of a modulated field is to deliver it to a film at right angles to the main axis of the radiation with some build-up material in front and to compare the irradiation of the film with the planned modulation. Another technique is to put the films on the head of the linear accelerator and do much the same. The use of integrating portal imaging has already been mentioned and perhaps the most important quality-assurance technique in IMRT in the clinic is the repeated mimicking of the Bortfeld-Boyer experiment, particularly in the early days of setting up IMRT in a particular institution.

5.   The techniques to deliver IMRT without a MLC

There are several techniques which can deliver IMRT using equipment which is not a conventional multileaf collimator. The first of these is tomotherapy. Tomotherapy techniques were independently proposed by two groups and can be categorised into two independent categories. The first to be described is that of the NOMOS Corporation and was announced, very unexpectedly, in 1992.

In this technique individual slices are separately irradiated and, in order to cover a tumour of finite volume, the patient must be longitudinally translated through the irradiation field. The second tomotherapy technique is that emanating from the University of Wisconsin, first discussed also in 1992, but not becoming a clinical reality until 2002 and in which the radiation performs a helical trajectory with respect to the frame of reference of the patient.

First we should discuss the NOMOS technique. The basis for the NOMOS technique is the multivane intensity-modulating collimator (MIMiC) which is a special collimator retro-fitted to the accelerator head. The radiation is collimated to a narrow slit and into this slit aperture can enter, for variable times, two sets of 20 stubby leaves or vanes which reside in two banks. The leaf thickness is about 8 cm and the leaves are moved by electropneumatic operation. Low-pressure air normally keeps the vanes closed. Turning a valve on sends

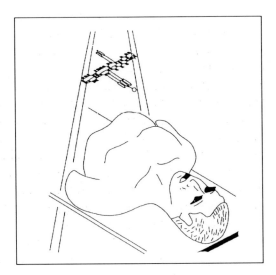

Figure 8. Shows the location of the MIMiC stubby-vane collimator in relation to the patient with some of the vanes closed and others open. The two back-to-back slit apertures each comprise 20 vanes.

high-pressure air to the front of the vanes to open them. The operation of the vanes takes 40–60 ms. This engineering would therefore fail safe if the air supply were to rupture. The leaf movement is activated by an on-board computer and the leaf patterns can change every 5°. In practice the leaves vary their dwell time in the aperture over each 5° arc interval and they do so in steps of 10% of the open-field intensity. Thus the spatial modulation is obtained through temporal variations. The leaves are focused in the transaxial direction; they are 6 mm wide on the patient's side and 5 mm on the source side. There is a five-element tongue-and-groove shape to prevent inter-leaf leakage. The aperture for each bank may be either 5 or 10 mm. The manufacturer claimed that the equipment could fit the linac of any other manufacturer and sensors monitor the leaf location. There is no direct connection between the equipment and the accelerator and inclinometers were used to measure the gantry orientation. The on-board computer, as well as functioning as a QA device, by which each leaf pair could be open or shut for inspection and QA tests, also contained a floppy-disk drive into which a floppy disk containing the leaf motion prescriptions was inserted. This floppy disk was created from a planning system known initially as PEACOCK and later as CORVUS. The tools and equipment manufactured by the NOMOS Corporation were all named after birds because of the hobby and interest of the managing director, Mark Carol.

The initial inverse-planning system inside PEACOCKPLAN was based on optimising a quadratic cost function by simulated annealing. Dose limits were imposed on structures and also importance factors were assigned to structures to balance the aggression of obtaining high tumour dose with the sparing of organs at risk. When this equipment was first invented and introduced into the clinic in the early 1990s, batch-mode calculations were required on independent computers overnight, not a specific limitation given that, in those days, this was the only commercial inverse-planning equipment available. Later developments used dose-volume constraints on the CORVUS system. The planning concepts were said by the Company to have been based on the work at the Royal Marsden Hospital and the Institute of Cancer Research by the author but it should be stressed that the development of the commercial planning system was performed independently by the company, the author's papers being in the public domain.

It came as a major surprise to the radiation therapy community when the MIMiC system was announced. The first European announcement was at the World Health Organisation on October 20th 1992 in Geneva at a meeting on 3-dimensional radiation therapy. The author had been invited to review inverse-planning mechanisms and, following him, came the lecture by Mark Carol. The author had confidently stated that inverse-planning algorithms did not have to take account of any specific constraints on delivering equipment because there was none. The next lecture gave the lie to this. The lecture was a sensational success and Mark Carol was given a second slot in the programme to amplify on his major invention. With hindsight it is known that the invention of the MIMiC is actually vested with the University of Wisconsin Group and in particular Stuart Swerdloff and Rock Mackie who had conceptualised the MIMiC and to whom royalties for its use were paid. However there is no denying that it was Mark Carol and the NOMOS Corporation in the early 1990s who got this technology rapidly into the clinic and the use of the NOMOS MIMiC dominated American IMRT delivery from the first treatments in Spring 1994 through to late 1997 time when commercial use of the multi-leaf collimator techniques began to compete. It is not totally understood why the technique did not have such an impact in Europe and, to the author's knowledge, only a handful of systems were installed whereas several hundred were installed in the United States of America. Towards the end of the 1990s the NOMOS Corporation introduced a two-dimensional MIMiC technique based on a chess-board of permanently attenuating cells and variably-attenuating cells in which mercury was fed into rubber balloons to create the variable attenuation. In order to generate a two-dimensional modulation, two irradiations were required with the whole apparatus rotated by 90°. This seems to have been a commercial failure and was never introduced for clinical use. As time has gone by the use of the NOMOS MIMiC has declined and it would not be appropriate in a teaching article to speculate on the reasons why.

Competing with the static Tomotherapy system throughout the 1990s was the development of the device in Wisconsin. The prototype was constructed based on a General Electric Advantage high-speed CT gantry and this accommodated an in-line linac and a dynamic MIMiC-like collimator with 64 vanes each projecting to 6.2 mm at 85 mm isocentre. It was described by Mackie as "a radiotherapy system in a box". The slit width was between 1 and 5 cm and the modified collimator had just one bank instead of two but with alternative leaves entering from each side to reduce friction and to allow the close packing of motors.

In the early experimental days, a virtual Tomotherapy system was created by putting a conventional NOMOS MIMiC on the head of linear accelerator and arranging that phantoms moved in a spiral longitudinal fashion by use of a simple kitchen chair with a wood spiral screw. The prototype Tomotherapy system was introduced into the clinic in the early 2000s and the first clinical treatment was in August of 2002.

There is no denying that there was a huge difference of approach between the two companies. The development work at NOMOS was totally out of the public eye and burst upon the clinical scene; the work at Wisconsin conversely took place in an academic organisation with papers being produced for over a

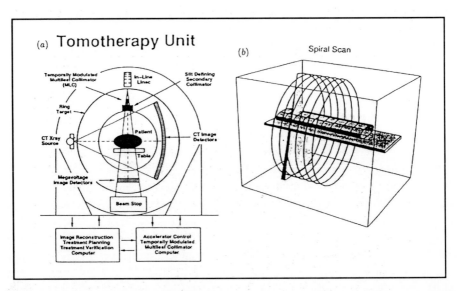

Figure 9. A schematic diagram of Mackie's tomotherapy. (a) An in-line linac with a temporally modulated MLC is mounted on a gantry along with other components which will perform megavoltage CT and diagnostic CT. The equipment rotates continually as the patient is translated slowly through the beam, hence (b) the motion is spiral with respect to the patient. (NB Publisher needs permission from Medical Physics for this one.)

decade on developments before finally reaching the clinical arena. At the time of writing many tens of Tomotherapy systems have been installed worldwide and there is an interesting debate taking place as to the merits of this system compared with the use of conventional multileaf collimators on conventional linacs.

Another technique for delivering IMRT without the use of the conventional MLC, indeed without the use of conventional linear accelerator, is robotic IMRT. The robotic radiosurgery system used to be known as the Neurotron-1000 and later became known as the Cyberknife. The system has gone through many embodiments but essentially the technique has remained the same, namely that of an in-line linear accelerator held on a commercial robot with 6 degrees of freedom of movement. The system has proved popular in selected American centres and also in the Far East from where much venture capital funding originated in the early development years. The system is marketed by the Accuray Company. What makes the technique particularly appropriate to IMRT of moving tumours is its ability to cope with intrafraction motion and that topic will be taken up in the next section.

Finally and very briefly we may remark on a number of alternative IMRT techniques that have reached only conceptual or prototype stage. These are those based on the use of cobalt irradiation, the use of jaws-plus-mask technique, the variable-aperture collimator, and techniques involving the direct placement of single-bixel attenuators.

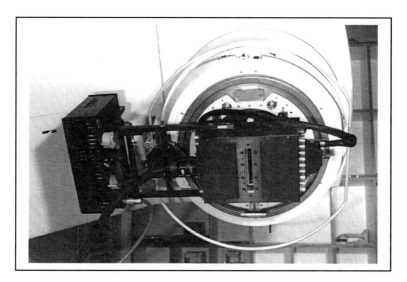

Figure 10. A view of the NOMOS MIMiC collimator attached to an Elekta accelerator at the Royal Marsden Hospital, showing a leaf pattern (Loan – courtesy of the NOMOS Corporation).

6. Techniques to compensate for organ-tissue motion

It might be reasonably said that the delivery of IMRT to the assumed static patient is already suitably optimised based on corresponding planning techniques. The current research agenda focuses now on the delivery of IMRT to the moving patient in which tissues, both target tumour and organs at risk, move. It was mentioned above that the Cyberknife provides a particularly convenient way of compensating for patient motion. Imagine that the patient has markers inserted in the target that can be visualised by external x-rays. The Cyberknife system is equipped with a pair of x-ray tubes and facing x-ray detectors. This radiation cannot be on all the time because it would overirradiate the patient but measurements can be made every so many (for example 10) seconds. At the same time the patient wears a set of infrared emitting markers on the skin surface and these are observed continually in real time by an infrared tracking system. Provided a correlation can be suitably established between the movement of the external markers and the movement of the internal markers a pseudo-continuous measurement of internal motion can be obtained. By feeding this knowledge back into the robot, the robot can automatically correct for the motion of tissues. It must be said that this is subject to a number of limitations. Firstly there is the difficulty of establishing the aforementioned correspondence. Secondly, the correspondence may vary and so new data are continually required and the model is updated on a first-in, first-out basis. Additionally, such measures generally can only take account of assumedly rigid-body motion and elastic-motion correction can never be formally and correctly accounted for using tracking techniques whether by this one or any other.

Intrafraction motion can also be corrected by transferring the known patient/organ motion on to the leaves of a multileaf collimator. The multileaf collimator is then made to "breathe" such that in the frame of the breathing collimator the moving tissue is stationary. This technique was first proposed in 2001 and has been subject to a large number of research studies looking at the many features and implications, in particular taking account of the density effect, the lack of rigid-body motion, the need to overcome the latency between making a measurement and using that measurement for correction and other factors. It may be observed that this has become one of the fastest growing research areas of endeavour in radiation therapy physics.

Alternatives to tracking moving tissue include the gating of a linear accelerator. This gating also requires to be "driven" by the observation of preferably internal markers and the radiation is switched on for only those periods where the tissue is considered to be in the planned treatment position. An alternative to this form of gating is the use of the active breathing co-ordinator

(ABC) device in which the breathing of a patient is regularly interrupted after a practice session to ensure that the patient is comfortable. Gating techniques by definition have a duty cycle of less than 100% but are relatively practical to implement.

Other techniques for correcting intrafraction motion include the deep-inspiration- breath-hold method, popular with some, but unpopular with others on the basis that it is less reproducible.

The whole of field image-guided radiation therapy (IGRT) is mushrooming. Image-guided radiotherapy includes all those radiotherapy techniques that rely on the use of (generally 3-D or 4-D) images to solve a variety of problems. At the planning stage IGRT can assist with disease staging and determining the gross tumour volume and the clinical tumour volume. It can assist assessing changes of target and the need to replan and it can assist adaptive therapy. Before each treatment fraction IGRT can help to position the patient correctly; this is so-called interfraction motion correction. During each fraction IGRT can assist with intrafraction motion correction by the methods already mentioned and post-therapy IGRT can assist with the assessment of response. A whole armoury of imaging techniques are employed for these tasks. They include the mundane use of kilovoltage x-ray film, optical markers, megavoltage film and portal imaging. Currently available in some centres but not widely available are PET, SPECT, MRI, megavoltage CT, ultrasound, and kilovoltage CT and implanted x-ray markers. Under development in many centres are the use of magnetic markers, optical markers, kilovoltage stereo imaging and various forms of breathing control including the use of spirometery and pressure belts. Other lectures at the ESMP concentrate on this major topic of IGRT. It can now be said that imaging is more firmly linked to radiotherapy than it ever has been in the past.

7. Conclusions on IMRT delivery

In 1988 when inverse planning seriously began there was no IMRT delivery equipment except the metal compensator. Planning techniques essentially ran well ahead of delivery possibilities. In 1992 MIMiC slice tomotherapy became available. In 1994 the MSF-MLC and the DMLC method had been operated by a few centres in a research setting. By 2000 commercial MLC/linac manufacturers have made available MSF-MLC and DMLC techniques linked to inverse planning. Today many centres in Europe, USA and Asia regard IMRT as a clinical necessity. The clinical implementation requires the multiple skills of doctors, physicists, radiographers, engineers all working together. It is not yet quite turn-key technology. The newer technologies, the use of helical Tomotherapy and robotics are beginning to challenge the use of conventional linear accelerators but currently vastly are outnumbered. Whether these reach the point of clinical implementation that C-arm linacs with MLC will cease to be used is very much a question to debate and observe with the passage of time. We may note that some groups are working on simpler IMRT to meet a call from less well-off places and, sadly, we may note an anti-IMRT backlash from some die-hard people. This is based on the largely true observation that the Phase 3, Level 1 clinical evidence for the clinical efficacy of IMRT is very limited. Nevertheless the momentum of IMRT seems unstoppable and it is now widely requested by patients. Motion is the enemy! The IMRT problem is largely solved for the static patient and the IMRT-IGRT problem is now the real one to address.

Finally we may comment briefly on the impact of articles in newspapers about IMRT. The author's centre published a Press Release that led to several newspaper articles in March 2000, 6 months ahead of the first clinical implementation of IMRT at this centre in the UK. This had an energising affect on the research itself. It also led to an increased awareness amongst patients and clinical staff of the possibility and such advertising generally leads to a move towards increased funding including that voluntarily by the public. In a climate in which the use of radiation is generally regarded as a "bad thing" such newspaper articles also draw the public's attention to the fact that the peaceful uses of radiation have enormous benefit to society with cancer fundamentally being cured by radiotherapy. A potential down side of newspaper articles is that they can sometimes misrepresent the exact scientific truth, the need to get the articles out generally precluding any form of proof reading as one would do with a scientific paper. A second down side is that it can lead to a worried public requesting new technology to be used in anger at a time when it is only just becoming implemented, newspapers not surprisingly wanting to concentrate on novel techniques and novelty rarely corresponds to mature clinical implementation.

In this chapter the techniques for delivery IMRT have been reviewed and linked to the rationale for IMRT and inverse planning for IMRT. Specific historical chronology has been documented and some comment has been given on the development and the current position and the future position. Of necessity such a review article largely with words rather than equations, graphs and multiple diagrams requires supplementing by the reader of either more comprehensive review text such as the four books to which this article refers and ultimately to a thorough study of the research literature. The books comprise very comprehensive lists of such literature.

References*

1. Bortfeld T, Schmidt-Ullrich R, De Neve W and Wazer D E Image-guided IMRT. Berlin-Heidelberg. Springer; 2006. ISBN 10-3-540-20511-X and 13-978-3-540-20511-1.
2. Mayles P, Nahum A and Rosenwald J C Handbook of radiotherapy physics. New York/London. Taylor & Francis; 2007. ISBN 0-7503-0860-5 and 978-0-7503-0860-1.
3. Mundt A J and Roeske J C Intensity modulated radiation therapy – a clinical perspective. Hamilton/London. BC Decker Ltd; 2005. ISBN 1-55009-246-4.
4. Palta J R and Mackie T R IMRT-the state of the art. Madison, WI. Medical Physics Publishing; 2003. ISBN 1-930524-16-1.
5. Schlegel W and Mahr A 3D Conformal radiation therapy 2nd edition. Heidelberg. Springer; 2007. ISBN 978-3-540-71550-4.
6. Van Dyk J The modern technology of radiation oncology. Madison, WI. Medical Physics Publishing; 1999. ISBN 0-944838-38-3 and 0-944838-22-7.
7. Webb S The physics of three-dimensional radiation therapy – conformal radiotherapy, radio-surgery and treatment planning. Bristol. IOPP; 1993. ISBN 0-7503-0247-X and 0-7503-0254-2 Pbk.
8. Webb S The physics of conformal radiotherapy – advances in technology. Bristol. IOPP; 1997. ISBN 0 7503 0396 4 and 0 7503 0397 2.
9. Webb S Intensity modulated radiation therapy. Bristol. IOPP; 2000. ISBN 0 7503 0699 8.
10. Webb S Contemporary IMRT – developing physics and clinical implementation. Bristol. IOPP; 2004. ISBN 0 7503 1004 9.

Disclaimer

The material in this chapter is new in the sense that it is a fresh presentation with new text of the overview of the delivery of IMRT. It is generated for the purposes of complying with NATO funding arrangements. The author has, of course, produced many similar texts in other books and review papers.

---

* There is a huge number of papers on this topic and so I have confined to referring to the overview books in which long lists of such detailed references can be found.

PART III:

HADRONTHERAPY

HADRONTHERAPY: CANCER TREATMENT WITH PROTON AND CARBON BEAMS

UGO AMALDI[1] AND GERHARD KRAFT[2]
[1]*University of Milano Bicocca and TERA Foundation*
ugo.amaldi@cern.ch
[2]*GSI and Technical University, Darmstadt*

Abstract Sixty years ago accelerator pioneer Robert Wilson published the paper in which he proposed using protons for cancer therapy. The introduction of protontherapy has been very slow, but in the last 10 years the field is booming and five companies offer turn-key centres. Fully stripped ions leave much more energy in the nuclei of the traversed cells than protons of the same range and are thus effective in controlling radio-resistant tumours which cannot be controlled neither with X-rays nor with protons. Paying particular attention to the European contributions, this contribution shortly reviews the history and the developments of carbon ion therapy, a recent chapter of the "hadrontherapy" which covers also radiotherapy with proton and neutron beams.

Keywords: Hadrontherapy; protontherapy; ion therapy

1. Introduction

In developed countries every year around 40,000 inhabitants per million are diagnosed as having cancer, some 50% of whom are treated with high-energy photons, called usually "X rays" by radiation oncologists. Today there are almost 10,000 electron linear accelerators (linacs) worldwide which produce the needed X rays. These photon beams have replaced low-energy X-rays and *the gamma radiation from radioactive cobalt used previously because they deposit the "dose" (i.e. the energy per unit mass) at a larger depth (Fig. 1).*

Y. Lemoigne and A. Caner (eds.), *Radiotherapy and Brachytherapy*,
© Springer Science + Business Media B.V. 2009

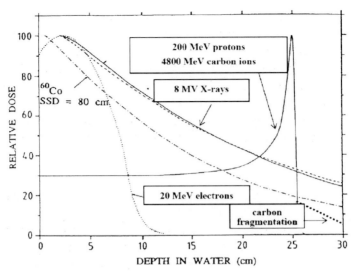

Figure 1. Depth dependence of the deposited dose for different radiations.

Modern linacs are practically identical to the first 3 GHz linear accelerator built in 1945 by William Hansen at Stanford University. The same year Robert R Wilson had been given by Ernest Lawrence the task of computing the thickness of the shielding for a 150 MeV cyclotron. Fifty years later, opening the conference "Advances in Hadron therapy" held at CERN in 1996, Wilson said: "I found that a few inches of lead would fix everything. But I did not stop. Why? Fifty years later I do not know why I did not stop. I suppose the first reason was just plain simple curiosity. So I went on and I jumped into the most obvious thing I could do next: because one could hurt people with protons, one could probably help them too. So I tried to work out every detail and I was surprised to see that the Bragg curve came up and came down very sharply".[1]

The narrow Bragg peak at the end of the range (Fig. 1) prompted him to publish a famous paper suggesting the use of protons to irradiate tumours while sparing – much better than with X-rays – the healthy tissue traversed, contiguous and located more deeply.[2] However, the "resonance" with the medical community was almost zero and it was a decade later before the first patients were treated at Berkeley and Harvard with proton beams produced by accelerators originally designed for experiments in nuclear physics. Eventually only in 1993 at the Loma Linda University Medical Center in California, the first proton synchrotron fully dedicated to proton therapy started to irradiate patients in three treatment rooms featuring magnets beam lines mounted on 10 m high "gantries", which rotate around the patient. It is no surprise that the Loma Linda synchrotron was built at Fermilab, the particle physics laboratory that Wilson created and then directed until 1987.

By 1993 about 15,000 patients worldwide had been treated with protons, which meanwhile had been proven to have the same biological and clinical effects as the X rays produced by a few mega electron volts linacs. By the beginning of 2007 the number has passed the 50,000 milestone. These patients have been irradiated in a dozen subatomic physics laboratories and in more than ten hospital-based proton therapy centres. Another ten centres are under construction or planned around the world since five companies now supply turn-key facilities. This single number justifies the statement that proton therapy is booming.

2.   Therapy with beam of carbon ions and number of potential patients

Heavier ions than protons, such as helium and later on argon, came into use at Berkeley in the 1957 and 1975, respectively. At the Bevalac, argon beams were tried in order to increase the effectiveness against hypoxic and otherwise radio-resistant tumours, i.e. tumours that need deposited doses two to three times higher if they are to be controlled with either photons or protons. But problems arose owing to non-tolerable side effects in the normal tissues. After a few irradiations, lighter ions, first silicon ions and then neon, were used for 433 patients until the Bevalac stopped operation in 1993.

The transition from protons to heavier ions adds another order of magnitude to the complexity of patient irradiation and a large number of radiobiological experiments on sub-cellular systems – such as DNA and chromosomes – and biological systems had to be performed. The research identified the systematic dependence of RBE on physical and biological parameters – mainly the capacity of cells to repair DNA damage – as the most important factor. In particular, the work showed that for beams of carbon ions the section of the particle track with increased RBE coincides with the few centimetres up to the Bragg peak, while for lighter ions it is concentrated only in the last few millimetres. For heavier ions, such as the argon, silicon, neon ions used previously at Berkeley, it causes significant damage in the normal tissues before the tumour.

The accumulated knowledge has been used by the GSI radiobiology group[3] to develop the Local Effect Model (LEM) which quantitatively derives the experimental data on the larger biological effectiveness of ion beams from the survival curves measured with X rays. At the beginning of 2008 this model has been for the more than 400 patients treated with carbon ions at the 'pilot project' proposed by one of us (GK) and built at the nuclear physics laboratory GSI in Darmstadt. A somewhat different approach has been used since 1994 at HIMAC, the Heavy Ion Medical Accelerator at Chiba in Japan which by the beginning of 2008 has treated about 4,000 patients.

While 200 MeV are needed to reach with protons deep-seated tumours (about 26 cm of water), 4,800 MeV (i.e. 400 MeV/u) are necessary for carbon ions, an energy which is 24 times larger. With carbon ions, the clinical results obtained in Japan and Germany on head and neck, lung, liver, and prostate tumours confirm the radiobiological predictions that they have a larger biological effectiveness than protons, because their ionization 24 times higher produces multiple double-strand breaks and clustered damages in the DNA of the traversed cell. These damages cannot be repaired by the usual cellular mechanism and, as a consequence, carbon ions are suited to control slowly growing tumours, which are precisely those tumours that are resistant to photons and protons. These types of tumours are thus the targets of choice in a carbon-ion facility, while proton therapy is well adapted to the cases in which a tumour is close to organs at risk that cannot be irradiated.[4]

Detailed analyses of the number of potential patients have been made in Austria, France, Germany and Italy.[5] The very consistent results of these different approaches show that about 1% of the patients treated today with X-rays should be irradiated with protons as the outcomes are definitely better than conventional therapy. In addition, about 11% of the X-ray patients would benefit from proton treatment but further studies are needed to quantify site by site the clinical advantages. Lastly, about 3% of the X-ray patients would benefit from carbon-ion therapy, but more clinical trials and dose escalation studies are necessary. Overall, 15% of the approximately 20,000 patients per ten million inhabitants treated with conventional radiation would receive a better treatment with hadron beams.

3. European centres for therapy with carbon ions and protons

In the past 5 years Europe has made important steps in the development and construction of hospital-based 'dual' centres for carbon ions and protons. Synchrotrons are at the heart of the European facilities since the magnetic rigidity of 400 MeV/u carbon ions is about three times larger than for 200 MeV protons, for which both cyclotrons and synchrotrons are in use. In future things will change since at present the Belgian company IBA is offering a novel 6 m diameter superconducting cyclotron that accelerates carbon ions up to 400 MeV/u.

Based on the successes of the GSI pilot project, the Heidelberg Ion Therapy Centre (HIT) designed by GSI was approved in 2001 and the civil engineering work began in November 2003. This centre features two horizontal beams and the first carbon-ion rotating gantry, which is 25 m long and weighs 600 t (Fig. 2). The first patient treatment will take place at the end of 2008.

At the end of 1995 one of us (UA), with the help of Meinhard Regler of the Med-Austron project, attracted the interest of the CERN management in the design of an optimized synchrotron for light-ion therapy. This was the starting point of a 5-year Proton and Ion Medical Machine Study (PIMMS).[6,7] As a development of this initiative, the TERA Foundation modified and adapted the PIMMS design and proposed it to the Italian government. In 2002 the project was financed by the Italian Health Minister. This is now being built in Pave (Fig. 3)

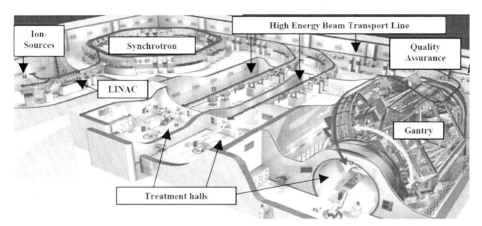

Figure 2. The delivery system of the Heidelberg Ion Therapy centre has been built by Siemens Medical Systems. As shown in the figure, the patient beds are mounted on robotic arms.

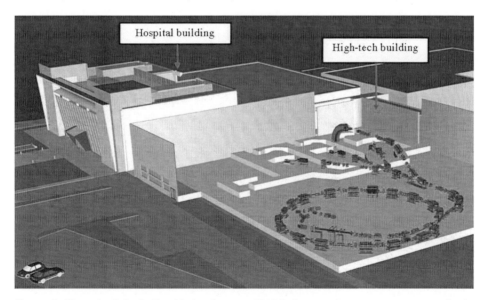

Figure 3. Phase 1 of the Italian National centre CNAO features three treatment rooms. In the central one, patients are irradiated with a horizontal and a vertical beam.[8] Two gantry rooms will be added in Phase 2 so that the centre will eventually feature five treatment rooms.

by the Centro Nazionale di Adroterapia Oncologica Foundation (CNAO) with strong construction involvement of INFN. The first patient will be treated at the end of 2009.

Recently other European dual projects have been moving forward. In 1998 the Med-Austron team presented the proposal of a dual centre to be built in Wiener Neustadt and eventually, in February 2007, the County of Lower Austria decided to build and operate MedAustron for a maximum investment of about €140 million. In 2007 MedAustron has signed an agreement with CNAO and obtained the construction drawings and the technical specifications of the centre being built in Pave.

In fall 1998 the University Claude Bernard of Lyon commissioned TERA a preliminary proposal of a hadrontherapy centre based on the PIMMS/TERA design. In March 2007 the Lyon ETOILE project was financed by the French and the regional governments. The tendering procedure will be completed by the end of 2008.

Moreover, in January 2006 contracts for a completely privately financed carbon/proton centre were signed by the Rhön-Klinikum-AG, which owns more than 40 German hospitals, including the Giessen-Marburg University clinics, and Siemens Particle Therapy (Fig. 4). The plan foresees that the centre will be completed by the end of 2010.

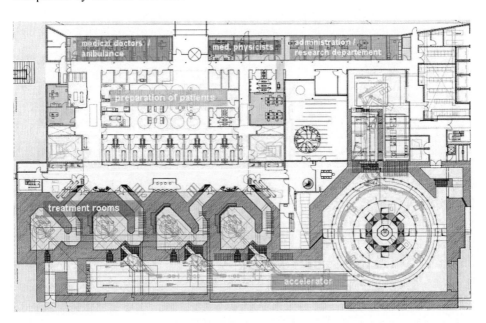

Figure 4. The design of the Marburg facility is based on an extended study of the clinical workflow. Three treatment areas with a horizontal beam line and one with a 45° oblique beam will be used to treat the patients.

All these activities have greatly profited from the many collaborations which have been set-up along the lines followed at CERN for the large physics experiments. Indeed, in a meeting held at CERN in February 2002 the five European projects (sited in Heidelberg, Pave, Wiener Neustadt, Lyon and Stockholm) decided to team with ESTRO (the European Society for Radiotherapy), EORTC (the European Organization for Cancer Research), CERN, GSI and TERA to form the European Network for Light Ion Therapy, which was financed for 3 years by the European Commission. The work done is documented in a series of reports can be found at the site www.estroweb.org/ESTRO/frame/template.cfm?id=90.

Between 2006 and 2007 a larger group of institutes and hospitals from 15 countries, coordinated by Manjit Dosanjh, joined forces and formed a 'platform' named ENLIGHT++

(see the site: http://enlight.web.cern.ch/enlight/).

The ENLIGHT++ platform has organized two European proposals, which have been approved with a total budget of more than ten millions: PARTNER, a Marie Curie Network, and ULICE (Union of Light Ion Centres in Europe), an Access program which is centered on the access of European patients and medical doctors to HIT and CNAO.

References

1. R R Wilson. 1997 Advances in Hadron therapy, U Amaldi, B Larsson and Y Lemoigne Eds, Elsevier.
2. R R Wilson. 1946 Radiology, 47, 487.
3. G Kraft. 2000 Prog. Part. Nucl. Phys. 45, 473–544.
4. U Amaldi and G Kraft. 2005 Rep. Prog. Phys. 68, 1861–1882. (This paper can be found at www.tera.it/ise/attach/DFILE/639/ROP.pdf.)
5. Carbon-ion therapy, Proceedings of the HPCBM and ENLIGHT meetings held in Baden (Sept 2002) and in Lyon (Oct 2003). 2004 Rad. Oncol. 73, Suppl. 2, 1–217.
6. L Badano et al. 1999 Proton-Ion Medical Machine Study (PIMMS). Part I CERN/PS 1999-010 DI.
7. L Badano et al. 2000 Proton-Ion Medical Machine Study (PIMMS). Part II CERN/PS, 2000-007 DR.
8. S Rossi. 2006 Proceedings of EPAC 06, 3631–3065.

PROTON THERAPY

UWE OELFKE
DKFZ, Heidelberg, Im Neuenheimer Feld 280,
69120 Heidelberg, Germany
u.oelfke@dkfz-heidelberg.de

Abstract Proton therapy is one of the most rapidly developing new treatment technologies in radiation oncology. This treatment approach has – after roughly 40 years of technical developments – reached a mature state that allows a widespread clinical application. We therefore review the basic physical and radiobiological properties of proton beams. The main physical aspect is the elemental dose distribution arising from an infinitely narrow proton pencil beam. This includes the physics of proton stopping powers and the concept of CSDA range. Furthermore, the process of multiple Coulomb scattering is discussed for the lateral dose distribution. Next, the basic terms for the description of radiobiological properties of proton beams like LET and RBE are briefly introduced. Finally, the main concepts of modern proton dose delivery concepts are introduced before the standard method of inverse treatment planning for hadron therapy is presented.

Keywords: Proton therapy; stopping power; LET; RBE; spot scanning; IMPT; treatment planning

1. Introduction

The main interest in radiation therapy with protons or heavy ions originated from its depth dose profile, the so called 'Bragg peak' shown in Fig. 1. Its most prominent feature is a maximum dose value at a penetration depth that can be controlled by the particles' energy. This peak is followed by a steep dose drop-off at a few millimeter increased depth, i.e., in contrast to conventional x-rays protons offer the advantage of a vanishing exit dose. Consequently, the overall dose burden to healthy tissue can be reduced by a factor of 2–3 if compared with standard radiotherapy practice.

Y. Lemoigne and A. Caner (eds.), *Radiotherapy and Brachytherapy,*   173
© Springer Science + Business Media B.V. 2009

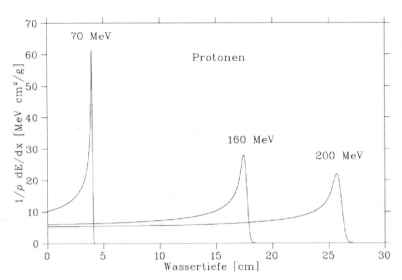

Figure 1. Proton depth dose curves in water for different kinetic energies. Mass stopping power in water versus penetration depth in water is shown.

Furthermore, the location of the 'Bragg Peak' in depth can be precisely controlled by the kinetic energy of the protons, i.e., this provides us with an additional degree of freedom for the realization of very tumor-conformal dose distributions. Another practical advantage of protons and heavy-ions is that the lateral position of the incident particle beams can be easily controlled by varying magnetic fields. This feature lead to the development of complex but very accurate and efficient dose delivery techniques like spot-scanning and raster-scanning.[1,2] The main anticipated advantage for heavy-ion beams are their radio-biological properties.

In the following we will briefly review the basic physics that determines the elemental proton dose distribution. Next, concepts for the assessment of additional radiobiological properties will be introduced. Finally, we will sketch the basic aspects of treatment planning related to proton dose delivery based on spot scanning techniques.

2. The proton pencil beam

Let us consider the case of an infinitely narrow beam of protons with kinetic energy E that enters a patient. The elemental dose distribution caused by this 'Pencil beam' is usually calculated for a patient consisting of the medium water. The 3-dimensional dose pattern factorizes in good approximation into a depth-dose curve[3] and a lateral distribution which describes the lateral scattering of protons from the central beam axis.[4] The physics of both components is presented in this section.

2.1. THE DEPTH DOSE CURVE: BRAGG PEAK

The depth dose curve is often defined as the dose measured along the central beam axis in a large treatment field. For a narrow pencil beam this is equivalent to a measurement where the total dose deposited in a well defined treatment depth is measured, i.e., all dose contributions from different lateral positions to one depth are accumulated.

The main energy loss of the protons occurs via Coulomb interactions with the electrons of the medium.[5] This electronic component of the interactions leads to an average energy loss per mean free path length, denoted as the stopping power $S_{el}(E)$, which is described by the well know Bethe Bloch equation. $S_{el}(E)$ is known be inversely proportional to the square of the velocity of the interacting proton, i.e., protons are gradually loosing all their kinetic energy and the rate of the energy loss is increasing the further the proton has entered the tissue. This explains the characteristic depth dose curve shown in Fig. 1. The position of the Bragg peak is close to the depth where the average energy of the protons has dropped to zero.

A second contribution to the proton depth dose curve arises from nuclear interactions, e.g. with oxygen nuclei. To first order these can be viewed as a proton absorption process, where the kinetic energy of the respective proton is locally absorbed. This dose contribution to the pencil beam is represented by the nuclear stopping power $S_{nuc}(E)$. Tabulated values of both proton stopping powers for various media can be found in the ICRU49.[5]

2.2. ENERGY AND RANGE

Closely related to the stopping power is the definition of particle range. Here the so called CSDA (Continuous slowing down approximation) range is universally employed. It is defined as:

$$R_{CSDA}(E) = \int_0^E dE' \frac{1}{S_{total}(E')} ,$$

with $S_{total}(E)$ as: $S_{total}(E) = S_{el}(E) + S_{nuc}(E)$. The value of a few characteristic range values of protons in water is indicated in Fig. 2. For protons the CSDA range is located at the depth of the 80% distal isodose curve, i.e. it is found at a slightly higher depth than the peak value.

A further direct consequence of the Bethe Bloch equation is that the ranges R(E,Z,A) of other heavy ion beams with mass number A, charge Z and the same energy per nucleon can be easily related to the respective proton ranges by:

$$R_{CSDA}(E; Z, A) = \frac{A}{Z^2} R_{CSDA}^{Proton}(E/A).$$

The energy dependence of the ranges is displayed in Fig. 3.

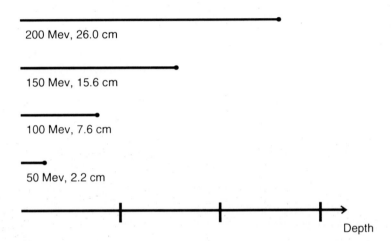

200 Mev, 26.0 cm

150 Mev, 15.6 cm

100 Mev, 7.6 cm

50 Mev, 2.2 cm

Depth

Figure 2. Characteristic values of CSDA proton ranges in water.

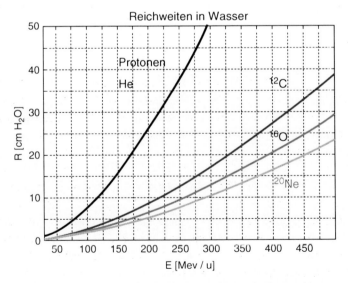

Figure 3. Energy range relationship for various heavy-ion beams in water.

2.3. THE LATERAL DOSE DISTRIBUTION

The lateral distribution of the dose of an elemental pencil beam again arises from Coulomb interactions with electrons and with protons of tissue composing nuclei. The net effect of beam broadening caused by Multiple Coulomb Scattering (MCS) is theoretically described by the more complex Moliere Theory.[4] However, it can be well approximated by a lateral dose profile of Gaussian shape where the width of the Gaussian curve is increasing with increasing penetration depth of the particles. The combination of the depth dose curve (displayed in Fig. 1) and the lateral dose spread through MCS finally results in a elemental pencil beam of proton dose as shown in Fig. 4.

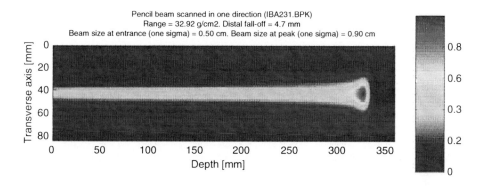

Figure 4. Relative dose distribution of a proton pencil beam in water. See color picture in Appendix I.

3. Radiobiological properties of protons

In contrast to conventional radiotherapy beams of photons and electrons, hadron beams usually are in addition characterized by an intrinsic radiobiological quality, i.e. the biological effect of these beams is not characterized by the commonly defined dose alone.[6] The intrinsic physical parameter responsible for this effect is the so called linear energy transfer (LET). For practical purposes it is often approximated by the stopping power $S_{total}(E)$. The LET accounts for the energy deposition on a far smaller spatial scale, i.e. on the nanometer to micrometer scale, and therefore refers to an individual property of the considered ion and its available kinetic energy. In general it is believed that a higher density of microscopic ionization events along a particle track causes an enhanced fraction of non repairable biological lesions in the tissue. Consequently, the observed biological damage observed per physical dose is higher for high LET radiation when compared to conventional low-LET beams like photons and electrons.

The increased potential of creating an enhanced biological effect with high LET radiation is quantitatively accounted for by the relative biological effectiveness (RBE). It is defined as the ratio of the dose of conventional photon beams (e.g., 60Co radiation) D_γ that causes a well defined biological effect to the dose D_p required by the particle beam to cause the same biological effect:

$$\text{RBE (biolog. system, biolog. effect)} = \frac{D_\gamma}{D_p}$$

For protons with LET-values usually ranging between 5–25 keV/μm[7] this extra radiobiological effect seems not be very pronounced. Particularly, all analyzed clinical data so far seem to indicated that a spatially invariant RBE value of 1.1 can account well for the radiobiological effects of proton beams.[8]

4. Dose delivery and treatment planning

4.1. DOSE DELIVERY

The extra degree of freedom provided by the energy and range selection of the proton beam combined with the possibility of magnetic lateral beam steering offers numerous methods for a tumour conforming dose delivery.

4.1.1. *Traditional passive beam delivery*

Traditionally, the energy and the lateral field width of the proton beam were determined by passive beam shaping devices,[9] i.e., a beam of fixed energy was intercepted by a set of beam modifiers such that a homogenous irradiation of the tumor target could be achieved from each selected beam direction. The depth dose curve was 'modulated in depth' by range modulator wheels. These were sweeping the Bragg peak of the highest energy over the projected volume of the target such that it was covered by a homogenous dose. The exact position of this Spread-out Bragg Peak (SOBP) was then determined by an additional range shifter. As an example a range modulator and an SOBP are displayed in Fig. 5.

Moreover, the distal dose gradient of each radiation beam can be tailored to the distal target edge by another patient specific absorber (bolus). The lateral treatment field is adapted to the projection of the tumor by a patient specific collimator.

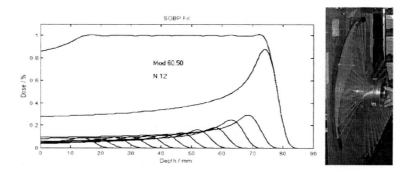

Figure 5. Spread out Bragg Peak (SOBP) (left-hand side) created by a range modulator wheel (right-hand side).

4.1.2. *Spot/raster scanning methods*

Almost all new proton therapy facilities, however, rely on modern spot-/or raster-scanning techniques for the dose delivery of proton beams. Here relatively narrow pencil beam of protons are laterally swept over the extension of the treatment volume. Combined with various modes of depth dose modulation, the following dose delivery modes are distinguished: (i) standard SOBPs, (ii) distal edge tracking, (iii) 2.5 D scanning where an SOBP is realized for each pencil beam, and (iv) the general 3D-scanning approach. These scanning methods are described in detail in reference.[10]

If a non-uniform tumor dose is delivered from each individual beam in such a way that the superposition from all treatment beams results in a homogeneous dose coverage of the tumor, the dose delivery is called 'intensity modulated particle therapy' (IMPT). Treatment planning of IMPT dose delivery is one of the most challenging optimization problems in radiation therapy.

4.2. TREATMENT PLANNING

Inverse treatment planning for proton therapy has to account for all potential dose delivery methods. Although, the calculation of physical doses for proton beams is relatively simple, there still remain enormous challenges to be dealt with.[11, 12] First, the number of optimization parameters – the fluence values for each single beam spot – can easily be in the order of $10.^4$ Moreover, the inclusion of local RBE-effects, specifically important for heavy-ion beams is

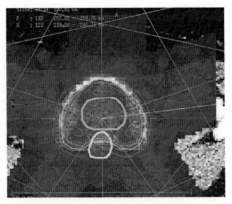

Figure 6. Treatment plan comparison for a prostate patient: left (IMRT photons); right (IMPT protons). On a transversal CT slice the dose distributions are shown as an overlay on the segmented anatomy of the patient. The shown volumes of interest refer to: prostate (light blue), extended target volume (red), bladder (dark blue) and rectum (yellow). See color picture in Appendix I.

still an area of active research.[13] Finally, a widely unsolved problem is the integration of treatment uncertainties in the optimization process or the consideration of organ movements.

Inverse treatment planning modules, however, are currently available and are applied in clinical practice. As a final example we show in Fig. 6 dose distributions on a central CT slice of a prostate patient, which were achieved for a treatment with five IMRT photon beams in comparison to proton therapy plan optimized for a distal edge tracking technique. The anticipated advantage of a significant reduction of medium and lower doses for the organs at risk, like rectum and bladder, is clearly visible.

References

1. Pedroni E., Bacher R., Blattmann H., Böhringer T., Coray A., Lomax A., Lin S., Munkel G., Scheib S., Schneider U. et al. The 200-MeV proton therapy project at the Paul Scherrer Institute: conceptual design and practical realization. Med Phys. 1995 Jan; 22(1):37–53.
2. Haberer T. et al., "Magnetic scanning system for heavy ion therapy", Nuclear Instruments and Methods in Physics, vol. A, No. 330, 1993, pp. 296–305.
3. Bortfeld T. An analytical approximation of the Bragg curve for therapeutic proton beams. Med Phys. 1997 Dec; 24(12):2024–33.
4. Gottschalk B. et al., Nucl. Instr. and Meth. B 1993; 74:467.
5. ICRU Report No. 49 Stopping Power and Ranges for Protons and Alpha Particles

6.   Kraft G. Tumor therapy with charged particles. Progr. Part. Nucl. Phys. 2000; 45:S473–S544.

7.   Wilkens J. J. and Oelfke U. Analytical linear energy transfer calculations for proton therapy. Med Phys. 2003 May; 30(5):806–15.

8.   Paganetti H., Niemierko A., Ancukiewicz M., Gerweck L. E., Goitein M., Loeffler J. S. and Suit H. D. Relative biological effectiveness (RBE) values for proton beam therapy. Int J Radiat Oncol Biol Phys. 2002 Jun 1; 53(2):407–21.

9.   Chu W. T., Ludewigt B. A. and Renner T. R. Instrumentation for treatment of cancer using proton and light-ion beams. Rev. Sci. Instrum. 1993; 64:2055–122.

10. Lomax A. Intensity modulation methods for proton radiotherapy. Phys. Med. Biol. 1999 Jan; 44(1):185–205.

11. Oelfke U. and Bortfeld T. Inverse planning for photon and proton beams. Med. Dosim. 2001 Summer; 26(2):113–24.

12. Oelfke U. and Bortfeld T. Optimization of physical dose distributions with hadron beams: comparing photon IMRT with IMPT. Technol Cancer Res. Treat. 2003 Oct; 2(5):401–12.

13. Wilkens J. J. and Oelfke U. Fast multifield optimization of the biological effect in ion therapy. Phys. Med. Biol. 2006 Jun 21; 51(12):3127–40.

PART IV:

BRACHYTHERAPY

PHOTON SOURCES FOR BRACHYTHERAPY

ALEX RIJNDERS
Europe Hospitals, Department of Radiotherapy, Av. de Fré,
1180 Brussels, Belgium
a.rijnders@europehospital.be

Abstract As introduction a short overview of the history of brachytherapy (BT) is given, with a focus on the evolution in the photon sources that have been used over the years. A major step in this evolution was the introduction of the automatic afterloading devices, which could be compared to the introduction of linear accelerators in external beam radiotherapy (EBRT). The modern afterloaders allow for optimization of the dose delivery and the use of different dose rates (low dose rate, high dose rate and pulsed dose rate) in function of tumor biology and patient comfort. Still today new sources are under investigation, and these developments together with the improvements in treatment planning and treatment techniques will enforce the role and place of BT as a valuable alternative for or supplementary to EBRT.

Keywords: Brachytherapy; photon sources

1. Introduction

It can be stated that the use of brachytherapy started with Röntgen's discovery of x-rays in 1895, and the first reported brachytherapy treatments for cancer were performed in 1903 in St. Petersburg using radium sources. Thus brachytherapy is as old as radiotherapy, and it certainly has not reached an end: over the years sources, equipment and techniques have kept on evolving, and today still new developments and important improvements are expected which will enforce the use of brachytherapy as treatment technique for cancer.

Y. Lemoigne and A. Caner (eds.), *Radiotherapy and Brachytherapy,*
© Springer Science + Business Media B.V. 2009

2.   History of brachytherapy sources

2.1. THE EARLY PERIOD

Initially (until the end of the 1950s when artificial isotopes became available) Radium-226 was the only isotope used, most often in the form of needle-like sources. Source strength was then expressed in mg Ra, and dosimetry systems and treatment reports usually referred to an amount of milligram-hours.

Already in the 1930s dosimetry systems were published which provided adequate rules to define how the sources had to be distributed in order to lead to a certain dose distribution over the treatment area. Most used was the Manchester System of Paterson and Parker, where the source distribution was adapted in order to produce a more uniform dose distribution. In the US the Quimby System was the system of choice, which adapted a uniform distribution of the sources and accepted the hot spots in the central area of the implant.

Asides the rather important dimensions of the radium needles the major drawbacks of the use of this isotope were linked with safety issues: ^{226}Ra decays into the gaseous ^{222}Rn, and it has to be ensured that the sources remain sealed to prevent radioactive radon leaking. Also the extremely long half life of ^{226}Ra (1,600 years) was a matter of concern with respect to radiation safety.

2.2. USE OF ARTIFICIAL ISOTOPES

At the end of the 1950s artificial isotopes such as ^{60}Co and ^{137}Cs became available which also found their way to be used in brachytherapy. Especially the latter became popular and caesium-needle sources were brought on the marked to replace the radium needles. An example of such a source is given in Fig. 1.

In a first instance the dosimetry systems were not modified, and source strengths were expressed as milligram-radium-equivalent in order to maintain the methodology.

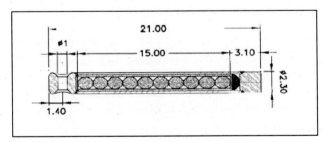

Figure 1. Example of a ^{137}Cs source as replacement of ^{226}Ra sources (Amersham model CDCS-M). The source contained a linear array of ten active beads of 1.3 mm in diameter borosilicate glass encapsulated in stainless-steel.

From the 1960s on different source models using ^{192}Ir as radioactive core were developed, and gradually ^{192}Ir became the commonly most used isotope for brachytherapy thanks to its favorable specific activity (allowing construction of small sources of adequate strength), acceptable half life (73.8 days) and median gamma ray energy (~400 keV).

This development led to the need of a new source strength definition, based on the *contained activity* of the source, the number of disintegrations per second. For practical dose calculations the concept of *apparent activity* was introduced, the activity that is emitted by the source taking into account absorption in the source itself and by its encapsulation. At first the unit Curie was introduced, where 1 Ci represents the activity of 1 g ^{226}Ra (3.7 × 10^{10} disintegrations/s). Later the Curie was replaced by the SI unit of Becquerel where 1 Bq = 1 disintegration/s. Today the use of apparent activity is no longer recommended for dose calculation.

2.3. AFTERLOADING TECHNIQUES

Radiation protection to the medical and nursing staff is an important issue in brachytherapy, and especially when the radiation oncologist has to manipulate and insert active sources directly into the patient he could be exposed to relative important doses.

Therefore already in the early years attempts were made to design manual afterloading techniques in which empty catheters were inserted in the organ or tissue that was to be treated and these catheters were only loaded with active sources after placement of all catheters and verification of the positioning.

Such manual afterloading techniques became generally more applicable from the 1960s onwards with the use of miniature caesium sources as shown in Fig. 2 and of iridium wire, hairpin or ribbon sources (technique for wire sources developed in Europe by the team from Institut Gustave Roussy in Paris, France).

Typical steps in a manual afterloading procedure are:

1. Careful placement of the applicators in the operating theatre
2. Radiographic or fluoroscopic imaging to verify and measure the position of the applicators, if these images are used for dosimetry then usually dummy sources are inserted to visualize the actual source positions
3. Calculation of the dose distribution and irradiation times using the registered images
4. Preparation of the sources: assembling of the combination of sources for intracavitary tube use or cutting of iridium wires/ribbons

5. Insertion of the radioactive sources in the applicators, the patient has been installed in a shielded room on a nursing ward
6. After completion of the treatment: removal of sources and applicators

Manual afterloading techniques resolved (partly) radiation protection issues towards the medical staff, but these issues still remained for the nursing staff and visitors, as treatment for radical treatments typically lasted for 4–5 days. Also the sources still had to be manipulated during source preparation, source insertion and removal.

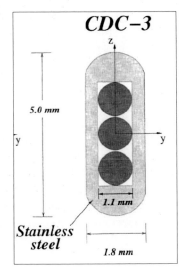

Figure 2. Example of a ^{137}Cs miniature source, the Amersham CDC-3. Three active beads (in the form of a borosilicate glass are contained in a welded stainless-steel capsule.

2.4. REMOTELY CONTROLLED AFTERLOADING DEVICES

The development of remotely controlled afterloading devices allowed for further reduction of the risk of radiation exposure to staff and visitors. Such devices enable sources to be transferred remotely from a shielded storage container in the device directly into the applicators and back.

Figure 3 shows a picture of one of the first devices, a prototype of the Curietron® which was used for treatment of gynaecological tumours using ^{137}Cs sources. While these first units only reduced the exposure due to the source preparation and insertion into the patient, later systems were developed which allowed the treatment to be interrupted and the sources to be retracted whenever medical or nursing staff had to enter the treatment room.

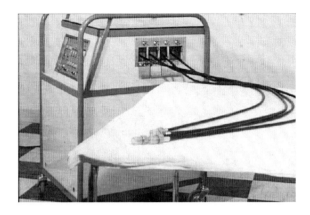

Figure 3. Photograph of the Curietron prototype, one of the first remotely controlled afterloading devices, used for the treatment of gynaecological tumours.

At present a large majority of brachytherapy treatments is performed using remotely controlled afterloading devices, allowing the patient to get medical and nursing attention and visits from close relatives in a safe and secure way, limiting the risk for irradiation of the staff to accidental situations. Treatment start and ending, and all interruptions are automatically logged by the system, and the system will control proper ending of the treatment. Safety systems (emergency and manual retraction) are foreseen to allow forced retraction of the source into the safe in case anything goes wrong.

3.   Dose rate in brachytherapy

Due to the limited total activity that could be used in a radium-application, traditional brachytherapy treatments were given in a continuous low dose rate (LDR) scheme. Also the radiation burden due to the manual manipulation of sources limited the maximum achievable dose rate.

When afterloading systems became available using caesium and iridium sources, and especially with the use of the remotely controlled afterloaders, dose rates could be increased and at present the majority of treatments is performed using high dose rate (HDR) techniques. ICRU[1] proposed the following definitions with respect to the dose rate: LDR 0.4–2.0 Gy/h, medium dose rate (MDR) 2–12 Gy/h and HDR > 0.2 Gy/min.

A special treatment schedule using remotely controlled afterloaders is the pulsed dose rate (PDR) treatment, in which a continuous LDR treatment is mimicked by administrating the dose in small steps (pulses), one pulse every hour or every 2 h, and adapting the dose given during each pulse so that the global average dose rate is in the LDR range (typically 0.6–1 Gy/h), i.e. the total treatment is spread over the same period as for an LDR treatment. The period between two pulses can be used for nursing the patient.

Figure 4 shows schematically how the dose is administered in the three different treatment schedules: in LDR a continuous low dose rate is used, in this example of 1 Gy/h. If this treatment would be performed using an PDR treatment unit, the same total dose would be given, but now in short pulses of 1 Gy given every hour. In the example the pulse dose would be given in 10 min, the next 50 min between two pulses the treatment is interrupted. When performing the treatment in an HDR schedule the total dose needs to be adapted taking into account the radiobiological effect of the higher dose rate and fraction dose. Depending on the total dose that needs to be given the dose would be administrated in a single fraction up to a few fractions, with sufficient time between fractions to allow for normal tissue repair.

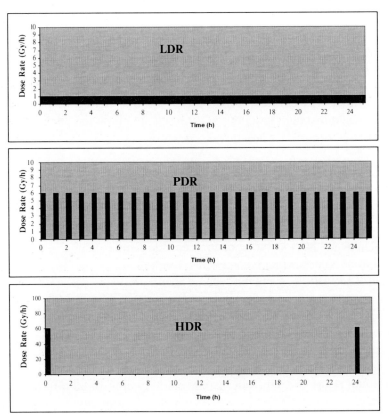

Figure 4. Schematic drawing indicating the way the dose is administered in three different treatment schedules: LDR (top), PDR (middle) and HDR (lower graph).

Another specific treatment schedule is the use of permanent implants, often used for prostate cancer treatment (permanent prostate brachytherapy PPBT). In this technique the radioactive sources (often ^{125}I or ^{103}Pd seeds) are implanted in the organ to be treated and remain there forever to decay completely. The total dose (typically 145–160 Gy for PPBT using ^{125}I) is then administered over a very long period (half life of ^{125}I is 59.49 days), while the dose rate decreases exponentially following the decay of the sources. For such permanent implants isotopes need to be used with an adequate half life and that emit photons of low energy: after implantation of the sources, the patient should be able to resume as soon as possible his normal way of life, without exposing his relatives and environment to severe radiation exposure risks.

4.   Commonly used sources in brachytherapy

As mentioned earlier the most frequently used isotope for brachytherapy sources at present is ^{192}Ir, especially as source for remotely controlled afterloading devices. Taking into account the half-life of 73.8 days, iridium sources need to be replaced every 3–6 months, depending also on how frequently they are used.

Afterloaders using caesium sources don't need to have the sources exchanged (^{137}Cs sources can be used for 10 years or more), and because of this advantage a number of caesium afterloaders are still in use for specific treatments such as gynecological applications. Caesium sources are also still available as tubes or pellets.

Less frequently used, Cobalt-60 brachytherapy sources are available as pellets or as HDR source for certain afterloading devices.

For permanent implantations (e.g. prostate brachytherapy) most often ^{125}I or ^{103}Pd seeds are used, while also ^{198}Au seeds are available.

TABLE 1. Physical characteristics of some photon sources used in BT.

Isotope		Average photon energy[a]	Half-life	Half value layer in lead (mm)
Cobalt-60	Co-60	1.25 MeV	5.26 years	10
Caesium-137	Cs-137	662 KeV	30.2 years	6.5
Iridium-192	Ir-192	380 KeV	73.8 days	6.0
Ytterbium-169	Yb-169	93 KeV	32.0 days	2.6
Thulium-170	Tu-170	84 KeV	128 days	0.25
Iodine-125	I-125	28.5 KeV	59.4 days	0.03
Palladium-103	Pd-103	20.8 KeV	17.0 days	0.02
Caesium-131	Cs-131	30.4 KeV	9.7 days	–

[a]Approximate values, depending on the source make and filtration.

Table 1 summarizes some physical characteristics for a number of commonly used or newly explored photon sources.

5.   Future developments

Manufacturers and clinical researchers are still investigating new isotopes to be used in brachytherapy. Several aspects can be taken into consideration such as photon energies and photon beam penetration in tissue and shielding materials (half-value layer in shielding material), the half-life of the source and the specific activity. Specific activity will determine the source dimensions and maximum available source strength, radiation protection requirements will depend on photon energies and half-value layer in shielding material, while half-life will be important for afterloader sources out of economical perspective or, in the case of permanent implants, for dose rate issues.

Some of the new isotopes that are studied by afterloader companies are for instance Thulium-170 or Ytterbium-169.

For permanent implants seed manufacturers continue investments in improving their products, other isotopes are also studied, and on the US market a company introduced Caesium-131 seeds to be used for PPBT, using the argument that the higher energy and shorter half-life of these seeds should be beneficial for the treatment of prostate cancer.

A very specific development that should be mentioned is the production of a miniature X-ray source (energy up to 50 KV), that could be used to replace the radioactive source of an afterloader.

In the recent years a lot of attention was put on enhancing the use of imaging modalities in BT treatment planning. While in the past dose prescription and reporting was often linked to fixed, geometrically defined rules, nowadays treatment planning and dose reporting based on accurately delineated target volumes and critical organ using appropriate imaging modalities such as CT, MRI or US is widely promoted. This evolution has also again pointed the attention on the issue of dose calculation accuracy: while in external beam treatment planning algorithms that can adequately handle dose calculation in a heterogeneous patient are now commonly available in commercial treatment planning systems, in BT most if not all of the systems still consider the patient to be a homogeneous, water-equivalent and infinite phantom. With the increased focus on the use of CT and MR data for dose calculation, many manufacturers are now also for their BT treatment planning system working on new algorithms, that can cope with the *real* patient's dimensions and tissue densities.

Along with these developments a lot of research is also focusing on the improvement of treatment techniques and the use of BT for new treatment sites. Therefore we can conclude that brachytherapy is still evolving, maybe with a little delay with respect to external beam radiotherapy (EBRT), but certain to play its role as an valuable alternative for or in addition to other treatment modalities such as surgery or EBRT for a number of treatment sites or indications.

In this chapter a general introduction was given into brachytherapy using photon sources. For more information on clinical aspects, background on dose prescription and reporting, and in depth elaboration of the physics aspects we can refer the reader to the reports and textbooks mentioned in the reference list.

References

1. ICRU, International Commission on Radiation Units and Measurements, Dose and volume specification for reporting intracavitary therapy in gynecology, ICRU Report 38, 1985.
2. ICRU, International Commission on Radiation Units and Measurements, Dose and volume specification for reporting interstitial therapy, ICRU Report 58, 1997.
3. The GEC ESTRO Handbook of Brachytherapy, Editors: A. Gerbaulet, R. Pötter, J.J. Mazeron, E. van Limbergen, available at the ESTRO website: *www.estro.org.*
4. A Practical Guide to Quality Control of Brachytherapy Equipment, ESTRO Physics Booklet No. 8, Editors: J. Venselaar, J. Pérez-Calatayud, available at the ESTRO website: *www.estro.org.*
5. R. Nath, L.L. Anderson, J.A. Meli, A.J. Olch, J.A. Stitt, and J.F. Williamson, Code of practice for brachytherapy physics: Report of the AAPM Radiation Therapy Committee Task Group No. 56. Med. Phys. 24, 1557–1598, 1997.
6. Radiation Oncology Physics: A Handbook for Teachers and Students, Editor: E.B. Podgorsak, available at the IAEA website: www.iaea.org.
7. Dosimétrie en Curiethérapie, A. Dutreix, G. Marinello and A. Wambersie, Masson, Paris, 1982.
8. The Physics of Modern Brachytherapy for Oncology, D. Baltas, L. Sakelliou and N. Zamboglou, Taylor & Francis, New York/London, 2007.

CALIBRATION OF PHOTON SOURCES FOR BRACHYTHERAPY

ALEX RIJNDERS
Europe Hospitals, Department of Radiotherapy, Av. de Fré,
1180 Brussels, Belgium
a.rijnders@europehospital.be

Abstract Source calibration has to be considered an essential part of the quality assurance program in a brachytherapy department. Not only it will ensure that the source strength value used for dose calculation agrees within some pre-determined limits to the value stated on the source certificate, but also it will ensure traceability to international standards. At present calibration is most often still given in terms of reference air kerma rate, although calibration in terms of absorbed dose to water would be closer to the users interest. It can be expected that in a near future several standard laboratories will be able to offer this latter service, and dosimetry protocols will have to be adapted in this way. In-air measurement using ionization chambers (e.g. a Baldwin–Farmer ionization chamber for ^{192}Ir high dose rate HDR or pulsed dose rate PDR sources) is still considered the method of choice for high energy source calibration, but because of their ease of use and reliability well type chambers are becoming more popular and are nowadays often recommended as the standard equipment. For low energy sources well type chambers are in practice the only equipment available for calibration. Care should be taken that the chamber is calibrated at the standard laboratory for the same source type and model as used in the clinic, and using the same measurement conditions and setup. Several standard laboratories have difficulties to provide these calibration facilities, especially for the low energy seed sources (^{125}I and ^{103}Pd). Should a user not be able to obtain properly cali-brated equipment to verify the brachytherapy sources used in his department, then at least for sources that are replaced on a regular basis, a consistency check program should be set up to ensure a minimal level of quality control before these sources are used for patient treatment.

Keywords: Brachytherapy; source calibration; reference air kerma rate; well type chamber

Y. Lemoigne and A. Caner (eds.), *Radiotherapy and Brachytherapy,*
© Springer Science + Business Media B.V. 2009

1. Introduction

In all protocols and guidelines on quality assurance for brachytherapy it is been recognized that the calibration of brachytherapy sources in the hospital is an essential part of a well-designed QA program. The medical physicist is in general responsible for the dose calculated for the patient, the source strength used for the dose calculation should therefore be validated independently by him, despite the fact that the manufacturer and/or company preparing the source(s) have determined and certified the source strength.

During the total chain of source production, calibration at the manufacturer's site, documentation and transport to the customer, many actions must be taken and the possibility of errors (mostly human) is not negligible.

The aim of the calibration is on one hand to ensure that the value determined by the local physicist and entered into the treatment planning system agrees with the source calibration certificate to within some predetermined limits. On the other hand the calibration should ensure traceability to international standards. This traceability is essential for comparison of treatment results at a national and international level.

In many countries calibration by the medical physicist in the hospital is a legal obligation, or imbedded in national protocols. At least it should be considered as a rule of good practice, being recommended in all international guidelines.

Several methods can be used to measure the strength of a brachytherapy source, using either a so-called "in-air" measurement technique, a well type ionisation chamber or a dedicated solid phantom. The choice of a method will depend amongst others on the source type (high energy/low energy, photon or beta ray) and the source presentation (seed/wire/needle/afterloader source).

The present chapter is based on the recommendations described in the IAEA TecDoc-1274[1] on the calibration procedures of photon and beta ray sources used in brachytherapy, and on the procedures described in ESTRO physics booklet No. 8.[2]

1.1. AIR KERMA RATE

At present the recommended quantity for the specification of the gamma sources is the reference air kerma rate, K_R defined by the ICRU Reports 38 and 58[3, 4] as the kerma rate to air, in air, at a reference distance of 1 m, corrected for air attenuation and scattering.

For needles, tubes and other similarly rigid sources, the reference point should be taken at right angles to the long axis of the source.

The SI unit of reference air kerma rate is Gy s^{-1} but as more practical units µGy h^{-1} for Low Dose Rate (LDR) sources is used, and µGy s^{-1} or mGy h^{-1} for High Dose Rate (HDR) sources.

The AAPM[5] recommends the use of a specific unit: the Reference Air Kerma Strength S_k, which is specified in terms of air kerma rate, $\dot{K}_\delta(d)$, *in vacuo* and due to photons of energy greater than δ, at distance d, multiplied by the square of this distance, d^2:

$$S_K = \dot{K}_\delta(d) \bullet d^2 \qquad (1)$$

At the reference distance of 1 m, $S_K = \dot{K}_R$.

1.2. CALIBRATION IN TERMS OF DOSE TO WATER

Although easy, fast and reliable methods exist for source calibration in terms of Air Kerma, it should be noted that the quantity of interest for the clinical user is the absorbed dose. At present calibration of chambers in terms of dose to water is not commonly available yet at the standard laboratories for the sources used in brachytherapy. However several projects are started to develop this methodology further, and it can be expected that in future source calibration directly in terms of dose to water will become the standard.

1.3. APPARENT ACTIVITY

Quantities such as equivalent mass of radium and apparent activity, A_{app}, are considered obsolete and no longer recommended for the specification of brachytherapy photon sources. However, source manufacturers still often use A_{app} on their source calibration certificates for source strength specification. Older brachytherapy treatment planning systems could also impose the use of it.
The apparent activity is related to the reference air kerma rate by:

$$A_{app} = \frac{d_{ref}^2 \bullet \dot{K}_R}{(\Gamma_\delta)_K} \qquad (2)$$

where $(\Gamma_\delta)_K$ is the air kerma rate constant, which will depend on the source construction, its encapsulation and on the photon energy. The problem is that a for many sources a number of values for the air kerma rate constant have been published. Using different conversion factors at the level of the manufacturer, the user or in the treatment planning system can cause confusion and unnecessary errors in dose delivery. The use of A_{app} should therefore be withdrawn.

2. In-air measurement technique

In-air measurement using an ionisation chamber is in most protocols considered the "Golden Standard" for the calibration of a high-energy photon source. The method cannot be used for low-energy sources such as ^{125}I or ^{103}Pd for several reasons:

- The uncertainty in the air kerma calibration factor for an air cavity chamber at these low photon energies is unacceptably high.
- In general, low energy photon sources have a limited reference air kerma rate. Thus the low measurement signal combined with a possibly relatively high leakage current will yield in measurements subject to a large uncertainty.

In this chapter we will concentrate on the calibration of ^{192}Ir afterloader sources. However, the correction factors given have only minor energy dependence and can therefore be used, without loss of accuracy, in calibration of ^{60}Co and ^{137}Cs brachytherapy sources.

2.1. RECOMMENDED IONISATION CHAMBERS

For the calibration of HDR sources, ionisation chambers with volumes greater than 0.5 cm^3 can be used (e.g. Baldwin–Farmer 0.6 cm^3 chamber). When calibrating LDR sources however, larger volume ionisation (up to 1,000 cm^3) may be needed to obtain a sufficient signal.

2.2. GENERAL FORMULA FOR AIR KERMA RATE CALCULATION

The reference air kerma rate $\dot{K}_R$ is specified at the reference distance of 1 m. The direct measurement at 1 m, however, is not always practical due to low signals and consequently possibly relatively high leakage currents of the ionisation chambers used. The reference air kerma rate, $\dot{K}_R$, may be determined from measurements made in-air using the equation:

$$\dot{K}_R = N_K \bullet (M_u/t) \bullet k_{air} \bullet k_{scatt} \bullet k_n \bullet (d/d_{ref})^2 \qquad (3)$$

where:
- N_K is the air kerma calibration factor of the ionisation chamber at the actual photon energy.
- M_u is the measured charge collected during the time t and corrected for ambient temperature and pressure, recombination losses and transit effects during source transfer in the case of afterloading systems.
- k_{air} is the correction for attenuation of the primary photons by the air between the source and the chamber.

- k_{scatt} is the correction for scattered radiation from the walls, floor, measurement set-up, air, etc.
- k_n is the non-uniformity correction factor, accounting for the non-uniform electron fluency within the air cavity.
- d is the measurement distance, i.e. the distance between the centre of the source and the centre of the ionisation chamber.
- d_{ref} is the reference distance of 1 m.

With equation (2) the reference air kerma rate is calculated on the day of measurement, source decay should be taken into account if the reference air kerma rate on another day is required.

2.3. AIR KERMA CALIBRATION FACTOR

The air kerma calibration factor, N_K, is obtained from a Secondary Standards Dosimetry Laboratory (SSDL) or directly from a Primary Standards Dosimetry Laboratory (PSDL). At present however, no primary standards exist for ^{192}Ir HDR calibrations. The air kerma calibration factor is determined by calibrating the chamber for an number of photon qualities and then using either an interpolation procedure or by polynomial fitting between the obtained factors.[6–8]

An alternative methodology can be that the standard laboratory calibrates a brachytherapy source of the same type and model and then uses this calibrated source to calibrate the users measurement equipment (this methodology is also used for the calibration of well type chambers, see further).

2.3.1. Measurement distance

A number of uncertainties in the measurement are influenced by the measurement distances, but in opposing directions:

- The uncertainty in the non-uniformity correction factor (effect of chamber size) decreases with increasing measurement distance.
- The positional uncertainty decreases with increasing measurement distance, as it follows the inverse square law.
- The influence of scatter increases with increasing distance, as the measurement signal will be reduced and thus scatter will be relatively more important.
- Similarly the effect of leakage increases with increasing distance as it will also become relatively more important.

The optimum measurement distance can be determined by minimizing the combined uncertainty due to the above effects. As example, the optimum distance has been shown to be 16 cm for a combination of an ^{192}Ir HDR source and a Farmer-type chamber.[9]

It should be noted that the non-uniformity correction factors used in most protocols are valid only when assuming point source geometry. The same assumption is needed as in equation (2) the inverse square relation is used. Thus the measurement distance must be large enough so that the source can be considered as a point source. As a rule of thumb the measurement distance be at least ten times the length of the source (error due to the point source approximation less than 0.1%). The following distances can be advised:

- For ^{192}Ir or ^{60}Co HDR sources: 10 cm $\leq$ d $\leq$ 40 cm
- For ^{137}Cs LDR sources, using large volume chambers: 50 cm $\leq$ d $\leq$ 100 cm

It is often recommended that the reference air kerma rate should be determined by doing measurements at multiple distances. The different values for $\dot{K}_R$ obtained at each distance should agree within measurement uncertainty, large variations can indicate problems in the experimental conditions and should be investigated.

Mechanical devices to ensure an accurate and reproducible measurement setup as shown in figure 2.1 can be used, but care should be taken that these do not cause to much additional scatter.

Figure 1. Example of a calibration jig suitable to calibrate an ^{192}Ir afterloader source (HDR and PDR). The source is transferred in the catheters 10 cm left and right from a centrally placed Farmer type ionisation chamber, and the readings for both catheters are averaged to correct for positional inaccuracies. A small plastic ruler assures that the catheters are kept 20 cm apart.

2.3.2. *Scatter correction*

In order to reduce the uncertainties, it is preferable to try to minimize the contribution of scattered radiation. This can be done by always using the same measurement setup and placing it (source and chamber) in the centre of the room, well above the floor (at least 1 m from any wall or floor).

The scatter correction can be determined by two different methods: the multiple distance method[6] and the shadow shield method.[10–12]

In the shadow shield method, a cone of a high Z material is placed between the source and the chamber in order to block out the primary photon beam, so the chamber will only measure scattered radiation. The scatter correction factor can then be calculated from the ratio of the measurement with and without the block in place. The cone must be thick enough to attenuate the primary beam sufficiently but should not be placed too close to the chamber to eliminate possible scattering from the cone. The advantage of this method is that the scatter component which is the quantity of interest, is directly measured. But measurements are difficult to carry out at typical calibration distances, e.g. between 10 and 50 cm due to the size of the cone and the distance to be kept from the chamber.

In the multiple distance method, measurements are performed over a range of distances (e.g. between 10 and 50 cm). The method assumes that the amount of scatter is constant over the range of measurements, which can be accepted to be valid for such a limited range of distances. After all other corrections the measurements are correlated with the inverse square of the measuring distances and the deviations from the inverse square law are then assumed to be due to the contribution of the scattered radiation.. The advantage of the method is that it is relatively simple to use and seems to agree quite well with measured scatter corrections.

In ^{192}Ir dosimetry it has been shown that the scatter correction factors obtained with the two methods are in good agreement.[10,11] An example of measured scattered correction factors using the shadow shield method is given in Table 1.

TABLE 1. Scatter correction factors determined with the shadow shield method at 1 m distance from an ^{192}Ir source.

Author	k_{scatt}	Chambers	Room size m × m × m
[10]	0.940	NE 2551 and Exradin A6	4 × 4 × 4
[10]	0.975	PTW LS-10	4 × 4 × 4
Petersen et al. 1994	0.940	Exradin A5	6 × 6 × 3
[11]	0.940	Exradin A5 and NE 2530/1	3.5 × 5 × 3.5
[12]	0.928	Exradin A4	–
[12]	0.941	Exradin A6	–

2.3.3. *Correction for the attenuation of primary photons in air*

In Table 2 correction factors k_{air} are given that correct for the attenuation of the primary photons between the source and the ionisation chamber, for different isotopes and in function of the measurement distance.[10, 11, 13]

TABLE 2. Correction factors for air attenuation of the primary photons from ^{192}Ir, ^{137}Cs and ^{60}Co brachytherapy sources.

Distance (cm)	^{192}Ir	^{137}Cs	^{60}Co
10	1.001	1.001	1.001
20	1.002	1.002	1.001
30	1.004	1.003	1.002
40	1.005	1.004	1.003
50	1.006	1.005	1.003
60	1.007	1.006	1.004
70	1.009	1.007	1.005
80	1.010	1.008	1.005
90	1.011	1.009	1.006
100	1.012	1.009	1.007

2.3.4. *Non-uniformity correction*

During in-air brachytherapy source calibration the ionisation chamber is exposed to a high divergent broad photon beam. This is quite different from the geometry used during the chamber calibration in collimated photon beams, and as the conditions that are present during brachytherapy source calibrations are not maintained, the calibration will not include the effect of divergence of the photons.

The secondary electrons entering the air cavity are mainly generated when the photons interact with the inner wall of the ionisation chamber. But due to the non-uniform photon fluency over the wall, the generation of secondary electrons from the wall will vary significantly from place to place in the wall. Thus the electron fluency in the air cavity of the chamber will also be non-uniformly distributed. The non-uniformity correction factor, k_n will account for this non-uniformity.

This factor depends on:

- The shape and dimensions of the ionisation chamber
- Material in the inner wall of the chamber
- The measurement distance and the source geometry ('point source', line source, etc.)
- The photon energy

Examples of non-uniformity correction factors are given in Table 3 for ^{192}Ir, but as variation in function of energy is minimal (less the 1.002) these values can also be used for ^{60}Co and ^{137}Cs sources.

TABLE 3. Non-uniformity correction factors, k_n, for some commonly used Farmer type ionization chambers in function of distance, valid for ^{192}Ir sources.

Chamber model	10.0	15.0	20.0	25.0	30.0	40.0	50.0
				k_n			
Capintec 0.65 cm^3 PR-06C Farmer	1.011	1.007	1.004	1.003	1.002	1.001	1.001
Capintec 0.6 cm^3 PR-05P AAPM	1.012	1.007	1.005	1.003	1.002	1.002	1.001
Exradin 0.5 cm^3 A2	1.001	1.001	1.001	1.001	1.001	1.000	1.000
Exradin 0.5 cm^3 P2	1.001	1.001	1.001	1.001	1.001	1.000	1.000
Exradin 0.5 cm^3 T2	1.001	1.001	1.001	1.001	1.001	1.000	1.000
Exradin 0.65 cm^3 Farmer A 12	1.012	1.007	1.005	1.003	1.002	1.001	1.001
NE 0.6 cm^3 Farmer, models 2505, 2505/A, 2505/3A, 2505/3B	1.011	1.007	1.005	1.003	1.002	1.001	1.001
NE 0.6 cm^3 Farmer 2571	1.009	1.005	1.004	1.003	1.002	1.002	1.001
NE 0.6 cm^3 Farmer 2581	1.009	1.005	1.004	1.003	1.002	1.002	1.001
PTW 1.0 cm^3 23 331 rigid	1.011	1.007	1.005	1.004	1.003	1.002	1.001
PTW 0.6 cm^3 Farmer 30 001	1.011	1.007	1.005	1.003	1.002	1.002	1.001
PTW 0.6 cm^3 Farmer 30 002	1.011	1.007	1.004	1.003	1.002	1.002	1.001

(Distance (cm) spans the 10.0–50.0 columns.)

2.3.5. *Correction for transit effects, leakage current and recombination losses*

When performing a measurement, the ionisation chamber will detect a signal as soon as the source leaves the shielding container. This signal measured while the source moves into the measurement position, and, after the measurement, again away into the storage container is called the transit signal. The magnitude strongly depends on the source-to-detector distance, and is significant at the distances used in calibration. It can be compensated by several techniques:

- Use of an externally-triggered electrometer which allows to perform the measurement after the source has stopped moving.
- Use a current reading after the source has stopped moving (if the signal is large enough).
- Subtracting two measurements taken over different intervals, the transit charge common to both can thus be eliminated.

The electrical leakage currents in the measurement system are more important then in external beam measurements as the signal levels in brachytherapy are quite low (typically 50–100 times less). It should be evaluated, and generally if the leakage is greater than 0.1% of the signal, it should be taken into account.

Correction for the recombination losses and for the ambient temperature and pressure follow the same principles as those used for external photon and electron beam dosimetry and are not further discussed here.

2.3.6. *Calibration using well type chambers*

Well type chambers offer an easy, reproducible and reliable method for calibrating brachytherapy sources. Condition for this is that provisions are available to ensure a reproducible positioning of the source within the chamber. In most commercial well type chambers a guide tube (insert) is provided to hold the source catheter along the axis of the cylindrical well.

The calibration point of a well type chamber is defined as the point at which the centre of the source is positioned during the calibration procedure; this point may differ from one source to another depending on the source length. If inserts are used during source calibrations in the hospital, the same should also be used when calibrating the well type chamber in the standard laboratory.

The sensitivity of the chamber will vary in function of the source position along long axis (so called sweet spot). This variation should be checked by varying the position of a small source along the length of the guide tube. Normally, over a trajectory of a few centimeters of the guide tube the signal will stay within some 1%, and the measurement point should be chosen in the centre of this maximum region. The location of the calibration point must be stated on the chamber's calibration certificate.

The well type chamber for brachytherapy source calibrations should be of a type designed especially for radiotherapy applications and preferably capable of measuring the reference air kerma rate for both LDR and HDR sources. Usually these chambers are open to the atmosphere and a correction for temperature and pressure has to be applied.

The use of pressurized well type ionisation chambers used in Nuclear Medicine Departments is not recommended for brachytherapy measurements because of the following reasons:

- The chambers generally measure only in units of activity.
- They have settings for given radionuclides but not for brachytherapy sources.
- The gas may leak from the pressurized volume which will cause the response to change over time.

- The thick walls required for the pressurization may absorb a significant part of the radiation to be measured, this results in a high-energy dependence.
- Without close control, the general use of the chamber may result in contamination from nuclear medicine procedures.

Measurements should always be done in the same manner and assuring a minimum scatter environment, with the chamber at least 1 m from any wall or floor. Special attention should be paid to temperature stabilization: the chamber should be left to come to equilibrium with its surroundings before beginning calibration. The minimal time necessary for this is 30 min as the temperature inside the well type chamber will slowly adjust to the room temperature. Direct measurement of the temperature of the air inside the well type chamber may be difficult, if not impossible, so care should be taken to let the chamber adjust to the room temperature.

A point of concern is the possible drift in the response of a well type chamber over prolonged periods of time, and it is generally recommended to use a long-lived source, e.g. a ^{137}Cs source, as a reference. By assuring reproducible positioning of such a source using a uniquely defined insert, one can verify the stability of the instrument's reading. Any sudden deviation of more than 0.5% in the check reading might indicate a problem. If the check source corrected reading remains within 2% of that at the time of the initial calibration, the assays may proceed but any possible reason for the deviation should be investigated. An alternative method for checking the chamber's stability is to irradiate it in an external ^{60}Co or linac beam under reproducible conditions.

Recombination corrections may be required if sources with high air kerma rate are used. Well type chambers may produce high ionisation currents with such sources, requiring correction for the recombination losses. Well known methodology using high and low collecting voltages to the chamber may reveal this effect and assessing its influence should be part of the acceptance procedure for the chamber. The verification should be repeated for lower activity sources, e.g. after one half life of an ^{192}Ir source. Typical 'high' and 'low' voltages that have been applied are 300 and 150 V. Note, however, that correction for recombination losses are necessary only if a correction for the chamber's collection efficiency was made during the chamber calibration and thus is included in the chamber's calibration factor.

The reference air kerma rate can be calculated from the measurements, using:

$$\dot{K}_R = N_{KR,u} \bullet (M/t) \tag{4}$$

where $N_{KR,u}$ is the well type chamber's calibration factor in terms of reference air kerma rate for the specific source type and model, M is the averaged charge measured during the time t and corrected for temperature, pressure and recombination loss.

It should be noted that well type chambers with thick internal walls may show an energy dependence which is particularly emphasized when calibrating low energy photon sources, such as ^{125}I and ^{103}Pd.

In general a well type chamber exhibits a larger dependence on the source design compared to Farmer type chambers. The well type chamber's calibration factor is valid only for the type of source it has been calibrated for. This is not only true for low energy photon emitters but also in ^{192}Ir HDR calibrations, i.e. a calibration factor that is valid for sources used in Nucletron HDR units may not be valid for other HDR units. In some cases, this dependence is stated on the well type chamber's calibration certificate but not always. In cases where there is no such statement, care should be taken in using the chamber under conditions different from the calibration conditions.

2.3.7. Calibration using solid phantoms

A solid phantom can be used as a tool in the source calibration process in two aspects:

- If the source has been calibrated by using the methods described earlier (in-air measurements techniques or a well type chamber), then a solid phantom may be used as a QC tool to check the calibration (e.g., the Meertens phantom,[14] figure 2.2).
- It can also be used for actual source calibration. In this case the phantom should be identical as described in the calibration protocol.

The big advantage of using a phantom is that the measurement distances are highly reproducible.

In the QA check, a measurement is made in a solid phantom and the ratio of the measured charge in the phantom to the charged measured during the calibration, should be constant from one source to another. As for the phantom measurements the distance is highly reproducible, it provides a simple method for monitoring the quality of the source calibration.

The other option is to use the phantom for the actual source calibration. Such commercially available phantoms are provided with suitable inserts for an ionisation chamber and for the source for the measurements, but for each specific phantom design correction factors needed to calculate the reference air kerma rate have be determined separately. For example DGMP gives generic correction factors for converting the measured charge or current into reference air kerma rate[15] when using the "Krieger" type phantom, a cylindrical PMMA phantom of well-described dimensions.

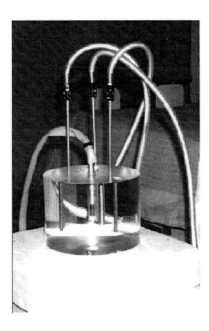

Figure 2. A solid PMMA phantom, suitable to receive the straight metal catheters of ^{137}Cs pellet sources (the "Meertens" phantom).

2.3.8. *Relative measurements*

It could be possible that in a given department a check of the source strength in terms of the absolute quantities, e.g. in mGy h^{-1} @ 1 m, is not possible. This could be due to the lack of traceability of the N_K factor for the specific sources or to the lack of resources within the physics department. Although this condition is highly undesirable, the physicist should try to develop a relative verification system with the locally available means.

It is recommended to have *at least* for source types that are replaced on a regular basis, i.e. the sources with a short half life, a *relative* system available to compare the results of own verification measurements with the value on the certificate supplied by the vendor of the sources. Consistency of readings of consecutive source deliveries can be thus be verified and any deviation larger than a certain level, for example 3% or 5% depending on the reliability of the system, should be further investigated. In this way serious incidents may be identified before the patient is treated.

The main issue is to have a stable set-up to avoid any uncertainty caused by variation in distance or positioning of the source, so the most reliable and reproducible system available should be used, for example a well type chamber or a measurement set-up in a solid phantom. The long-term stability of such a set-up can again be checked with a long lived source (e.g. ^{137}Cs).

References

1. IAEA, International Atomic Energy Agency. "Calibration of photon and beta ray sources used in brachytherapy. Guidelines on standardized procedures at Secondary Standards Dosimetry Laboratories (SSDLs) and hospitals". IAEA-TECDOC-1274, March 2002.
2. A Practical Guide to Quality Control of Brachytherapy Equipment, ESTRO Physics Booklet No. 8, Editors: Venselaar, J., Pérez-Calatayud, J. available at the ESTRO web site: *www.estro.org.* 2004.
3. ICRU, International Commission on Radiation Units and Measurements. "Dose and volume specification for reporting intracavitary therapy in gynecology". ICRU Report 38, 1985.
4. ICRU, International Commission on Radiation Units and Measurements. "Dose and volume specification for reporting interstitial therapy". ICRU Report 58, 1997.
5. Nath, R., Anderson, L.L., Meli, J.A., Olch, A.J., Stitt, J.A. and Williamson, J.F. "Code of practice for brachytherapy physics: Report of the AAPM Radiation Therapy Committee Task Group No. 56". Med. Phys. 24, 1557–1598, 1997.
6. Goetsch, S.J., Attix, F.H., Pearson, D.W. and Thomadsen, B.R. "Calibration of ^{192}Ir high-dose-rate afterloading systems". Med. Phys. 18, 462–467, 1991.
7. Borg, J., Kawrakow, I., Rogers, D.W.O. and Seuntjens, J. P. "Monte Carlo study of correction factors for Spencer–Attix cavity theory at photon energies at or above 100 keV". Med. Phys. 27, 1804–1813, 2000.
8. Van Dijk, E. "Comparison of two different methods to determine the air kerma calibration factor (Nk) for [192]Ir". In: Proceedings of the "International Symposium on Standards and Codes of Practice in Medical Radiation Physics". IAEA-CN-96-75, Vienna, Austria, 2002.
9. DeWerd, L.A., Ezzell, G.A. and Williamson, J.F. "Calibration principles and techniques". In: "High Dose Rate Brachytherapy: A Textbook". Library of Congress Cataloging-in-Publication Data, Editor: Nag, S., 1994.
10. Verhaegen, F., Van Dijk, E., Thierens, H., Aalbers, A. and Seuntjens, J. "Calibration of low activity ^{192}Ir sources in terms of reference air kerma rate with large volume spherical ionisation chambers". Phys. Med. Biol. 37, 2071–2082, 1992.
11. Drugge, N. "Determination of the Reference Air Kerma Rate for Clinical [192]Ir Sources". Thesis, Internal Report, University of Göteborg, 1995.
12. Piermattei, A. and Azario, L. "Applications of the Italian protocol for the calibration of brachytherapy sources". Phys. Med. Biol. 42, 1661–1669, 1997.
13. Palani Selvam, T., Govinda Rajan, K.N., Nagarajan, P.S., Bhatt, B.C. and Sethulakshmi, P. "Room scatter studies in the air kerma strength standardization of the Amersham CDCS-J-type 137Cs source: a Monte Carlo study". Phys. Med. Biol. 47, N113–N119, 2002.
14. Meertens, H. "In-phantom calibration of Selectron-LDR sources". Radiother. Oncol. 17, 369–378, 1990.
15. DGMP, Deutsche Gesellschaft für Medizinische Physik. Krieger, H. and Baltas, D. "Praktische Dosimetrie in der HDR-Brachytherapie". Report No. 13 of DGMP, 1999 (a) (in German).
16. Petersen, J.J, van Dick, E., Grimbergen, T.W.M and Aalbers, A.H.L. "Absolute determination of the reference air kerma rate for MicroSelectron-HDR 192 Ir source, serial number 098." Report S-E1-94.02, Utrecht (1994).

CANCER OF THE CERVIX: EXAMPLES OF BT TREATMENT TECHNIQUES

INGER-LENA LAMM
Lund University Hospital, Lund, Sweden
Inger-lena.lamm@skane.se

Abstract Brachytherapy treatment techniques for cancer of the cervix are presented, with examples based on the "Stockholm" technique.

Keywords: Brachytherapy; intracavitary applicators; reporting

1. Introduction

Brachytherapy, BT, has played and is still playing a major role in the radical treatment of cervical cancer, often combined with external beam radiotherapy, EBRT. Brachytherapy is also combined with surgery, both pre- and post-operatively. Today, "chemoradiotherapy" is often used for more advances cervical cancers, radiotherapy and chemotherapy are combined concurrently.

The earlier experience of BT treatments is based on the use of radium sources with manual loading of the sources and long treatment times, in the order of days, low dose rate (LDR) treatments. Several systems for intracavitary BT of cervical cancer were developed, e.g. the Stockholm, Paris and Manchester systems, specifying applicator design, source strength and source distribution, dose rate, dose specification and total treatment time. These systems have been extensively used with good clinical results, and constitute the basis for modern BT of cancer of the cervix (for a summary of these systems, see ICRU Report 38, Ref 1).

Development of manual, and later remote-controlled, afterloading equipment, and the use of ^{137}Cs, ^{60}Co and ^{192}Ir sources, gave new possibilities to optimize the treatment and also made it possible to minimise the radiation dose to staff, a problem with the manual radium techniques.

The introduction of high dose rate (HDR) remote-controlled afterloading equipment with a programmable "high activity" [192]Ir mini-source has opened the possibility to use individual optimisation of the dose distribution. Different types of applicators have been developed, both for LDR, HDR and pulsed dose rate (PDR) applications, e.g. ring applicators, Manchester- and Fletcher-type applicators, vaginal cylinders and moulds.

A true individualisation of the dose distribution requires a 3D definition of tumour, target and organs at risk, with the applicator in place. Consequently, modern commercial applicators for BT for cancer of the cervix are available also in versions compatible with modern imaging techniques, e.g. CT and MR.

For a detailed presentation of brachytherapy for cervical cancer, the reader is referred to the GEC-ESTRO Handbook of brachytherapy (Ref 5), and the GEC-ESTRO recommendations (Refs 2–4).

2. Prescribing, recording and reporting in radiotherapy

The International Commission on Radiation Units and Measurements, ICRU, has published a number of reports on dose specification and reporting for external beam radiotherapy. ICRU Report 62 "Prescribing, Recording and Reporting Photon Beam Therapy" (Supplement to ICRU Report 50) states: "When delivering a radiotherapy treatment, the volumes and the doses must be specified for several purposes: prescribing, recording and reporting. It is not the goal nor the task of the ICRU to recommend treatment techniques and absorbed dose levels. *Prescription* of a treatment is the responsibility of the radiation-oncology team in charge of the patient. *For reporting purposes*, it is important that clear, well-defined, unambiguous, and universally accepted concepts and terminology are used to ensure a common understanding. Only under these conditions can a useful exchange of information between different centers be achieved."

This statement is just as valid for brachytherapy as for external beam therapy. Further, in order to retain as much consistency as possible with dose and volume specification for external beam radiotherapy, it is desirable to use the same terms and concepts wherever possible in brachytherapy.

The ICRU has published two brachytherapy related reports: ICRU Report 38 (1985) "Dose and Volume Specification for Reporting Intracavitary Therapy in Gynecology" and ICRU Report 58 (1997) "Dose and Volume Specification for Reporting Interstitial Therapy". Today, the recommendations of both ICRU 38 and 58 are being developed and extended with the introduction of image-guided techniques also in brachytherapy.

ICRU 38 covers definition of basic terms and concepts used in intracavitary brachytherapy for gynaecological cancers, afterloading techniques, dose patterns, definitions of volumes and reference points, and recommendations for reporting. The "simple" ICRU bladder and rectum reference points, based on orthogonal radiographs, have been used extensively to characterize BT for cervix cancer in terms of maximum doses to these organs, in spite of their well known short comings. These ICRU reference points are still used for reporting purposes to relate image guided techniques to older techniques.

Both CT and MR imaging of patients with cervical cancer allow definition of bladder, rectum, bowel, vagina and sigmoid, while MR alone gives the possibility to define also the tumour volume. In the future, methods of prescribing and reporting BT for cancer of the cervix will be replaced by methods making use of 3D MR imaging with the applicator in place. Recommendations on image based 3D treatment planning in cervix cancer BT have been presented by the gynaecological GEC ESTRO working group, see Refs 2–4.

3. The old Stockholm system and the newer ring applicator, the traditional way

3.1. THE RADIUM SYSTEM – LDR MANUAL LOADING TECHNIQUE

The classical Stockholm system, a radium-based low dose rate intracavitary system with manual loading of the sources, used an applicator consisting of an intrauterine tube and a vaginal box. The modified system used in Lund had a connector sleeve, making it possible to connect the two parts of the applicator outside the patient, insert the whole applicator as a package and finally unfold the intrauterine probe perpendicular to the box with the applicator in place. Treatments were traditionally specified in "mg*h", the amount of radium used in milligrams and the treatment time in hours. (Total reference air kerma, TRAK, is the modern equivalent to "mg*h".) The geometrically defined "point A" (2 cm up and 2 cm out) was later used as specification point, the definition adopted from the Manchester system. The dose rate in both bladder and rectum was measured at the applicator insertion, and the prescribed dose lowered when readings were "high". (Note that depending on the pressure used on the measurement probe, you could change the reading dramatically due to the high dose rate gradients around the applicator.) Figure 1 shows a connected Lund radium applicator and a typical dose rate distribution, with the dose rate at point A indicated.

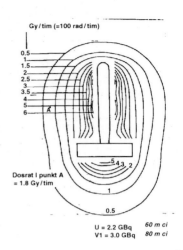

Figure 1. LDR radium applicator, the modified Stockholm system used in Lund, manual loading of the applicator. The intrauterine tube is securely coupled to the vaginal box, which is pushed against the cervix. A typical dose rate distribution is shown to the right.

3.2. THE NEW RING APPLICATOR – HDR REMOTE AFTERLODING TECHNIQUE

In the mid 1990s, the old LDR radium treatments were replaced by HDR treatments using a remote afterloading unit with a stepping ^{192}Ir source. The applicator was changed to the "Stockholm system" based ring applicator, consisting of an intrauterine probe and a ring, with the ring part securely fastened at right angles to the probe, see Fig. 2. This replacement implied many changes at the same time, changes in (a) applicator geometry, (b) applicator fixation, (c) dose and dose rate distribution around the applicator e.g. in the target volume, (d) dose specification, (e) dose to organs at risk, (f) fractionation schedule, (g) in the combination of EBRT and BT, and also changes in the EBRT doses. This change from a well established technique, with good treatment results, to a new technique had to lead to at least the same clinical results as before, i.e. to at least as good cure rates and at least as few complications as before.

The intracavitary BT target volume was in words defined as: (a) cervix including the primary tumour, (b) part of the corpus uteri, (c) part of the vaginal mucosa to 2 mm depth and (d) part of the proximal parametria. Standard type dose distributions were discussed and accepted for nominal target volumes, according to the above definition, for the available ring diameters and intrauterine probe lengths, see Fig. 3. Spacers were designed to displace organs at risk away from the high dose regions "below" the ring, see Fig. 2.

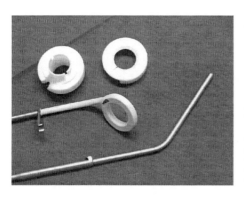
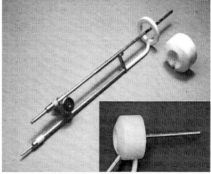

Figure 2. HDR remote afterloading technique, ^{192}Ir source. Ring applicators with plastic spacers; the ring and the intrauterine probe are securely locked in a well-defined position. Different ring diameters and intrauterine probe lengths are available. (GammaMed, Varian.)

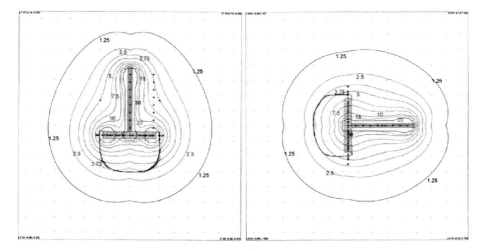

Figure 3. Standard dose distributions, coronal and sagittal planes, for the ring applicator with spacers designed to keep organs at risk outside the high dose region; diameter of ring with spacer – 4.1 cm, intrauterine probe length – 4 cm. Dose specification: 5 Gy = 100% at 2 mm depth in the vaginal mucosa ventrally and dorsally (the dose is 100% at point A for this ring–probe combination). Dose to organs at risk (rectum, bladder), nominally 5 mm outside the spacer is 77% = 3.85 Gy.

3.3. VERIFICATION OF SOURCE STOP POSITIONS IN THE RING APPLICATOR

CT and MR compatible ring applicators (titanium applicators, Varian) are also available, and a standard dose distribution based on a CT scan in water is shown in Fig. 4 for such a ring. The marker wires, simulating source stop positions, are clearly visible. Caution has to be taken when interpreting marker wire stops as true source stop positions in the ring applicator.

For a correct treatment delivery, it is necessary to verify both the correct applicator and source stop geometry and that the source stop positions in the applicator during treatment correspond to the stop positions used for treatment planning.

In the GammaMed remote afterloading unit (Varian), the source on its wire is first driven out quickly until it is fully extended, i.e. it reaches the distal end of the applicator. (All combinations of applicator and source guide tube, connected to the afterloader, are adjusted to exactly the same total length.) The treatment proper starts and the source stops, retracts, stops etc. according to the treatment plan.

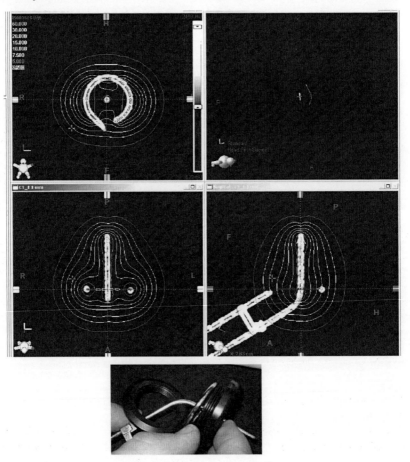

Figure 4. CT scan in a water tank of a CT/MR compatible ring applicator, titanium, with plastic spacers (Varian), three orthogonal planes. Also shown is a standard dose distribution. Note the marker wires inserted into the two channels, the intrauterine probe and the ring.

When the source is pushed out to its end position, the wire will follow the outer surface of the ring channel due to the natural tension in the source wire. The same situation applies for the marker wire, when it is pushed into the ring channel. However, when the source wire is retracted, the wire will follow the curvature of the inner ring channel surface instead, with the source tip sliding along the outer channel surface. The effect on the source movement will be two-fold; first, the source will not move at all with the first few mm of retraction, second, the effective source stop positions will not lie along the outer channel surface. Thus, the marker wire will not represent the true stop positions.

To find the true representation of the source stop positions in the ring channel, radiochromic film can be used, see Fig. 5. From the films, the diameter of the stop positions can be determined, as well as the initial retraction required to give the same "true" step size for the first step as for the following steps. Knowledge of the diameter of the stop positions is a basic requirement for correct dose distribution calculations!

The ring applicator is a rigid applicator, and the definition of treatment planning geometry should be part of the commissioning of the applicator.

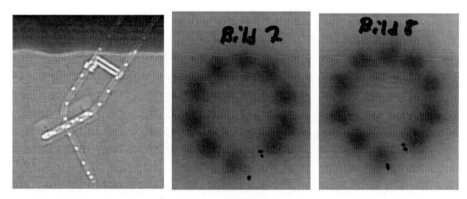

Figure 5. CT scan of the titanium ring applicator in a water tank, with marker wires inserted. The two radiochromic film images show the actual source stop positions in the ring channel using a step size of 5 mm (Varian, GammaMed); "Bild 7", treatment started "directly", result <5 mm between the two first stops, 5 mm step size for the following stops; "Bild 8", source retracted 2 mm before start of treatment, result 5 mm step size for all stop.

3.4. THE APPLICATION PROCEDURE

The patient is under anaesthesia during the application, which always starts with a gynaecological examination and an assessment of the present clinical situation. A bladder catheter with the balloon filled with contrast medium (7 cm^3) is used to identify the bladder risk point. The applicator most suitable for the patient is chosen and inserted; first the intrautrine probe, then the ring with its

spacers. Ultrasound is often used during the application procedure to quali-
tatively verify the applicator position, see Fig. 6. An indicator with lead-shot is
finally inserted in the rectum, to represent the rectal position. Orthogonal
radiographs are taken, see Fig. 7.

From the radiographs, reference point maximum doses to bladder and
rectum are identified. Radiobiological effects for the combination of EBRT and
BT are continuously calculated/estimated both for the organs at risk, bladder
and rectum, and for the target. The total treatment is planned to give as high a

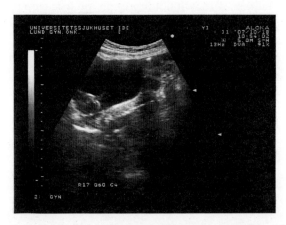

Figure 6. Ultrasound verification of applicator position; the intrauterine probe is clearly seed in
the uterus. A bladder catheter is used, with 7 cm^3 contrast in the balloon, to define the bladder
dose calculation points. Note, that the bladder is partially filled here to visualize the uterus; the
bladder is "empty" during treatment.

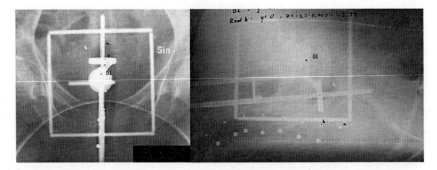

Figure 7. Orthogonal X-ray images; ring applicator including plastic spacers (not visible on the
images), diameter ring including spacer = 3.4 cm, intrauterine probe length = 4 cm, ring- and
probe-parts correctly connected, 7 ml contrast in Foley catheter balloon in bladder (Bl) Special
indicator with two rows of lead-shot in rectum (a, b). Silver marker in cervix (elongated – for
verification of appl. pos. between fractions). Calculations of total radiobiological effect (external
beam radiation and brachytherapy) for organs at risk (dose restrictions) and target are made.
Marker wires indicating source stop positions inserted in both probe and ring.

dose as possible to the target while keeping the dose to the organs at risk below accepted risk levels, formulated as dose limits, "action levels". The dose limits used for the risk organs are 72 Gy in 36 fractions for the rectum and slightly higher for the bladder (76–78 Gy in 38–39 fractions), based on clinical experience using this calculation method. BT doses are always discussed and adapted to the clinical situation individually for each patient at each BT fraction. The rectum and bladder point doses calculated in this way do not reflect the true maximum doses, as has been shown from studies using 3D imaging based treatment planning. Especially the bladder maximum dose is usually higher. A discussion of risk organ doses, maximum doses, doses to 1 cm^3, doses to 2 cm^3, and which dose measure that relates to different types of side effects etc can be found in Refs 2–4.

The traditional "system based" method of prescription, the treatment planning based on orthogonal radiographs and standard dose distributions give little possibility to individualise and adapt the dose distribution to tumour volume and to organs at risk. Prescription and treatment planning based on modern 3D imaging methods using CT and MR, perhaps also ultrasound, are required.

4. The future, using image guided brachytherapy, in 4D

Modern imaging techniques will make it possible to adapt each brachytherapy fraction for the individual patient. But, the possibilities are connected to a number of requirements. There are requirements on education and training in using and interpreting current imaging modalities, especially addressing oncology needs in defining tumour extent and target volumes. Further, an infrastructure is required, which allows imaging of the patient with the applicator in position followed by definition of structures, treatment planning and finally treatment in one session, with the applicator still in the same position in the patient.

Going from a standard type protocol for cervix cancer brachytherapy, from a target definition "in words", to a 3D definition of tumour, target and organs at risk, it is essential that you base the transition on your own experience, your own clinical results, as well as on recommendations and published results. CT imaging will allow definition of organs at risk, but MR imaging is recognized as the modality necessary to define tumour extent. CT imaging with the applicator in place will make it possible to evaluate risk organ doses for standard protocol plans, and as a first step towards individualization minimise doses to risk organs without compromising "the old type target dose". The full individualization, requiring MR based planning, and delineation of tumour, target and organs at risk, will allow simultaneous optimization of doses to all these structures. As the tumour shrinks during radiotherapy, 4D treatment planning should ideally be used, i.e. each BT fraction should be optimized in 3D. For larger tumours, the traditional intracavitary applicators will not be able to give dose distributions covering the target volume without giving rise to too large doses centrally in the target. To overcome this problem, ring applicators combined with interstitial needles have been developed.

As mentioned earlier, the introduction of modern imaging techniques will change the methods of prescribing, recording and reporting brachytherapy for cancer of the cervix. It is of great importance to use the same language when communicating clinical results and a common language is a prerequisite in the brachytherapy community for comparing and evaluating clinical results. The ICRU Report 38 has laid the foundation in 1985, and developments of recommendations for 3D image based treatment planning in cervix cancer brachytherapy have been presented by the GEC ESTRO working group, see Refs 2–4. Useful reading is also found in Ref 5.

References

1. ICRU Report 38: Dose and Volume Specification for Reporting Intracavitary Gynecology (1985).
2. C Haie-Meder et al. Recommendations from Gynaecological (GYN) GEC-ESTRO Working Group (I): concepts and terms in 3D image based 3D treatment planning in cervix cancer brachytherapy with emphasis on MRI assessment of GTV and CTV. Radiother Oncol 2005; 74:235–245.
3. R Pötter et al. Recommendations from gynaecological (GYN) GEC ESTRO working group (II): Concepts and terms in 3D image-based treatment planning in cervix cancer brachytherapy – 3D dose volume parameters and aspects of 3D image-based anatomy, radiation physics, radiobiology. Radiother Oncol 2006; 78:67–77.
4. S Lang et al. Intercomparison of treatment concepts for MR image assisted brachytherapy of cervical carcinoma based on GYN GEC-ESTRO recommendations. Radiother Oncol 2006; 78:185–193.
5. The GEC ESTRO Handbook of Brachytherapy, Editors: Alain Gerbaulet, Richard Pötter, Jean-Jacques Mazeron, Erik van Limbergen (2002), available at the ESTRO website: www.estro.be.

PARIS SYSTEM FOR INTERSTITIAL BRACHYTHERAPY

GINETTE MARINELLO
Unité de Radiophysique et Radioprotection, C.H.U. Henri Mondor, 51 avenue du Maréchal De Lattre de Tassigny, 94000 Créteil, France
ginette.marinello@hmn.aphp.fr

Abstract The Paris System is a complete dosimetric system which greatly facilitates brachytherapy, using iridium 192 sources at low, pulsed or high dose-rate. The aim of this chapter is to present briefly the part of the system dedicated to interstitial brachytherapy. After a short description of the sources which can be used, the three basic principles of the Paris System are presented together with its particular mode of dose specification within the implanted volume, and the fixed value of the Reference Isodose (RI) equal to 85% of the Basal Dose-rate (BD), representative of the arithmetic mean of the minimal dose-rates in the central region of the implant. The method to calculate the treatment time is given. Simple relationships which can be used to predict the minimal dimensions of the treated volume (volume encompassed by the RI) at the very moment of the implant are presented.

Keywords: Interstitial brachytherapy; Paris System; iridium 192; forecast dosimetry

1. Introduction

Currently used alone or associated with other modalities, interstitial brachytherapy is a very interesting technique for irradiating accessible tumour sites such as skin, breast, head and neck, anus and rectum, bladder, etc. When the patient anatomy permits, the simplest way to optimize the dose is to judiciously arrange the radioactive source guides throughout the target volume while following the rules of a predictive dosimetry system. The aim of this chapter is to present briefly a very complete system,[3] called the Paris System, including at one and the same time forecast dosimetry and recommendations for specifying the dose and calculating the treatment time.

For more detailed information and numerous examples of application, the reader can refer to "A Practical Manuel of Brachytherapy".[9]

2. Sources

Paris System is generally used with low activity sources of iridium 192 under the form of wires (straight lines or loops), single pins, hairpins and seed ribbons, or with the single moving source of low, pulsed or high dose rate afterloading devices.[1,3] In this last case, the system remains unchanged but the prescribed dose has to be modified to take into account the differences in radiobiological effect due to the dose rate:

$$^{HDR}DOSE = {}^{LDR}DOSE \times W_{rate}$$

When the Paris System is used with loops or hairpins (Marinello, Wilson et al. 1985, Pernot, Marinello et al. 1997), it is important to be sure that the branches are not too far apart and remain parallel over a sufficient length to obtain a correct dose distribution (Fig. 1A). In practice it may be assumed that a loop is correctly formed when the height of the curved part is less than, or equal to, half the spacing (d) between the branches, and when the length (b) of the branches is at least 1.5 times the spacing (Fig. 1B).

When the system is used with ^{192}Ir seed ribbons, it has been shown that the spacing between two consecutive seeds should be less than or equal to 1.5 times the active length of the seed (Marinello, Valero et al. 1985).

It can also be practiced using the single moving source of low, pulsed or high dose rate afterloading devices, provided that the steps and the dwell times be identical and the preceding conditions be fulfilled *(for instance: steps of 5 mm for a HDR source of 3.5 mm in length).*

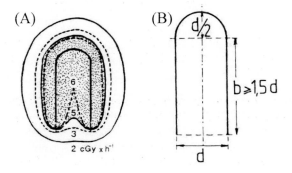

Figure 1. Loop correctly formed.

3. Principles of the Paris System

The Paris system is very flexible. The number, the arrangement and the spacing of the sources can be freely chosen provided that the parameters lead to a treated volume of correct dimensions and the three following principles be followed.

3.1. RADIOACTIVE SOURCES SHOULD BE PARALLEL AND ARRANGED SO THAT THEIR CENTER BE IN THE SAME PLANE, CALLED "CENTRAL PLANE"

The *central plane* should be perpendicular to the long direction of the sources if they are located at the same level and are similar in height (Fig. 2A, B), or in the mid-plane of the application if there are differences in height less than 1 cm between two nearby sources (Fig. 2C) or if the sources are curvilinear (Fig. 2D).

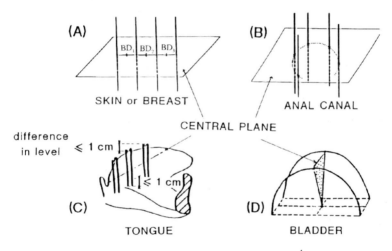

Figure 2. Definition of central plane.[4]

3.2. THE LINEAR REFERENCE KERMA RATE (OR LINEAR ACTIVITY) SHOULD BE UNIFORM ALONG EACH SOURCE AND BE THE SAME FOR ALL SOURCES

For seed ribbons, or stepping source with regular steps and dwell times, the mean linear reference kerma rate along the different trajectories has to be considered.

3.3. THE RADIOACTIVE SOURCES SHOULD BE EQUIDISTANT

When the thickness of the volume to be treated is thicker than 1–1.2 cm, two, or rather seldom three planes of sources have to be implanted. The principle of equidistance implies that the intersections of the sources with the central plane should be arranged following the apices of equilateral triangles (spacing between planes equal to 0.87 times the spacing between the sources), or the corners of squares (Fig. 3).

When prefabricated templates are used, it is essential to make sure that the above principle can be followed.

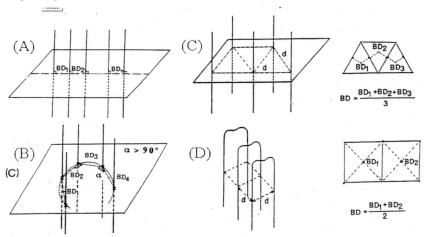

Figure 3. Definition of basal dose-rate BD in the central plane for planar implants (A), sources arranged in curved planes (B), in "triangle patterns" (C) or square patterns (D).

4. Specification of dose

The particularity of the Paris System is that the dose is specified within the target volume, and not at the boundary of the implant as for the other systems.[7,10]

4.1. BASAL DOSE-RATE

The judicious arrangement of radioactive sources as recommended by the Paris system leads to minimal variation of the dose in the region of the central plane, characterized by minima, called *elementary basal dose-rates (BD$_i$)*.

Therefore, the dose distribution is expressed as a percentage of the basal dose-rate (BD) which is the arithmetic mean of BD$_i$ within the implant:

$$BD = \Sigma \; BD_i/i$$

BD depends on the radionuclide used, the geometry of the implant and the linear air kerma rate of the sources. By convention, and in order to make easy the comparison between different implants, BD is expressed for 1 h and sources of linear reference air kerma rate of 1 μGy h^{-1} m^2 cm^{-1}. Then, the actual linear reference air kerma rate of the sources is taken into consideration for the calculation of treatment time (see section 3.3). The method remains valid for the somewhat uneven spacing often encountered in practice, provided that the homogeneity of the dose remains acceptable within the target volume. It has been shown[2] that it is always the case when the component DB$_i$ obeys the relationship:

$$0.90 \; BD \leq DB_i \leq 1.10 \; BD$$

4.2. REFERENCE ISODOSE

The dose is specified along an isodose surface of value equal to 85% of the basal dose-rate BD. This value has been justified by years of clinical experience and avoids subjective interpretation.

4.3. CALCULATION OF TREATMENT TIME

Calculation should be done after the implant is in place because the implant may not be exactly what was intended. Several manual or computerized methods that are useful for checking the quality of the implant and calculating BD[11] can be used. As the value of the reference dose-rate is 85% of BD, the treatment time (T) is calculated using the simple relationship:

$$T = [\text{prescribed dose}/(BD \times 0.85 \times K_R \text{ on the day of the implant})] + \Delta t$$

where Δt is a term taking into account the decay of the sources during the application when necessary (for instance implants performed with low dose-rate sources of ^{192}Ir).

5. Different volumes

The *treated volume* is the volume surrounded by the isodose surface of value 85% of BD. Its dimensions (length, thickness, lateral or safety margin) are defined by reference to minimal dimensions of the reference isodose measured within the implanted volume.[2]

The hyperdose volume is the volume surrounded by the isodose surface the value of which being the double the reference value (170% of BD).

6.   Forecast dosimetry

6.1.   PREDICTIVE RELATIONSHIPS

Constant relationships between the geometric implantation data and the dimensions of the treated volume have been established assuming that the principles of the Paris System are respected and that all sources in a given pattern have the same active length. Predictive relationships for rectilinear sources equal in length are summarized in Table 1. It has been shown that they are also valid, under certain conditions, for lines of slightly unequal length.[2]

Other simple relationships are published for straight sources arranged on a curved surface used for instance for anal canal implants, loops and hairpins currently used for head and neck implants (Marinello, Wilson *et al.* 1985, Pernot *et al.* 1997), etc.

TABLE 1. Predictive relationships for rectilinear sources equal in length.

Patterns	Treated length/active length	Treated thickness/spacing	Lateral margin/spacing	Security margin/spacing
Two lines	0.7	0.5	0.37	–
N lines in one	0.7	0.6	0.33	–
N lines in squares	0.7	1.25	–	0.27
N lines in	0.7	1.33	–	0.20

6.2.   HOW TO USE PREDICTIVE RELATIONSHIPS IN CLINICAL PRACTICE?

Using the Paris System, the number, the arrangement, and the spacing of the radioactive sources can be freely selected at the time of the application, provided the chosen parameters give a treated volume with appropriate dimensions. The different steps to be followed are summarized in Table 2.

In fact, clinical practice has revealed the following:

- The lower limit of spacing ranges from 8 mm for short lines up to 15 mm for longest lines (10 cm or more) for practical reasons related to the ease and accuracy of source implantation (parallelism of lines more difficult to obtain with narrow spacing and error of parallelism less significant with wider spacing).
- When the spacing between sources is excessive, namely more than 15 mm for short lines and 22 mm for long lines (10 cm or more), tissue necrosis are

observed in the immediate vicinity of the sources (size of hyperdose volumes). This occurs particularly with delivery of brachytherapy alone at doses of 60–70 Gy, or implantation performed in previously irradiated tissues.

TABLE 2. Paris System: forecast dosimetry.

Use of predictive relationships

Step 1	Carefully evaluate the target-volume to be implanted and define its three dimensions
Step 2	Determine the number of source planes to be used from the thickness: one plane, or two planes or more (rather seldom), if thickness >12 mm
Step 3	Considering the shape of the target volume, determine the form of the sources to be used and their arrangement (coplanar pattern or patterns "in triangles" or "in squares")
Step 4	Using the ratio "treated thickness/spacing", determine the minimal spacing between sources, and then the number of sources required
Step 5	Calculate the length of sources to be used from the ratio "treated length/active

References

1. Boutroux-Jaffre (1997), Photon-Emitting Sources. In "A Practical Manual of Brachytherapy, Medical Physics Publishing, (Madison, WI), 3–21.
2. Dutreix A., Marinello G. and Wambersie A. (1982), "Dosimétrie en curiethérapie", Masson (Paris), 109–161.
3. ICRU (1997), Dose and Volume Specification for Reporting Interstitial Therapy, Report No 58, Oxford University Press.
4. Marinello G. and Pierquin B. (1997), The Paris System, Optimization of dose, and Calculation of Treatment time. In "A Practical Manual of Brachytherapy", Medical Physics Publishing (Madison, WI), 53–68.
5. Marinello G., Valero M., Leung S. and Pierquin B. (1985), Comparative dosimetry between iridium wires and seed ribbons, *Int. J. Radiat. Oncol. Biol. Phys.* 11, 1733–1739.
6. Marinello G., Wilson J.F., Pierquin B., Barret C. and Mazeron J.J. (1985), The guide gutter or loop techniques of interstitial implantation and the Paris system of dosimetry, *Radiother. Oncol.* 4, 265–273.
7. Meredith W.J. (1957), "Radium Dosage: The Manchester System", Livingston Ed. (Edingburgh).
8. Pernot M. and Marinello G. (1997), Head and Neck. In "A Practical Manual of Brachytherapy", Medical Physics Publishing (Madison, WI), 127–146.
9. Pierquin B. and Marinello G. (1997), "A Practical Manual of Brachytherapy", Medical Physics Publishing (Madison, WI).
10. Quimby E.H. (1947), Radium dosage in radiumtherapy, *Am. J. Roent.*, 57, 622–627.
11. Rosenwald J.C. (1997), Calculation of dose distribution. In "A practical Manual of Brachytherapy", Medical Physics Publishing (Madison, WI), 29–51.

SOLID STATE DETECTORS FOR QUALITY ASSURANCE IN BRACHYTHERAPY

GINETTE MARINELLO
Unité de Radiophysique et Radioprotection, C.H.U. Henri Mondor, 51 avenue du Maréchal De Lattre de Tassigny, 94000 Créteil, France
ginette.marinello@hmn.aphp.fr

Abstract Because of the low energy of photon or beta radioactive sources currently used for modern techniques of brachytherapy, and the high dose gradient existing around them, the dosimeters to be used for quality controls of the sources or absorbed dose measurements should have special characteristics such as an accurate and reproducible response, a low dose response dependence as a function of energy, a high sensitivity under a small volume, a response independent of the dose-rate, etc. Commercial solid state detectors, such as diodes, MOSFETs, EPR-alanine, Gafchromic films, plastic scintillators and radiothermoluminescent (RTL) dosimeters, can be interesting in the concerned field. After a brief reminder of their principle, a review of their dosimetric advantages and disadvantages is presented. A selection of detectors the best suited for brachytherapy with examples of use for quality assurance controls or in-vivo dosimetry is also presented.

Keywords: Brachytherapy; solid state detectors; diodes; MOSFETs; EPR-alanine; Gafchromic films; plastic scintillators; TLD

1. Introduction

Many solid state detectors are well adapted to Quality Controls (Q.C.) or *in vivo* dosimetry (IVD) in brachytherapy (BT), because they have a large sensitivity under a small volume, most often without electrical connexion, and exhibit interesting dosimetric characteristics, such as an accurate and reproducible response, a fairly low or no-dependence of their response versus energy, dose-rate, etc. The selection of the commercial detectors suitable for BT measurements should be made considering the mass energy absorption coefficient of the detectors relative to water for photon source emitters, and the energy stopping powers coefficients of detectors relative to water for beta emitters, in the energy range interesting the BT. The corresponding data are shown in Figs. 1 and 2 for photon and beta emitters, respectively. They show that all detectors are suitable for dose measurements with photon emitters of energy ≥ 100 keV except argentic films and diodes, which can only be used to check source position or displacement. All detectors are suitable for dose measurements with beta emitters, except argentic films.

A brief review of the suitable detectors for brachytherapy is made in the following sections. Their advantages and disadvantages are pointed out, and examples of application given.

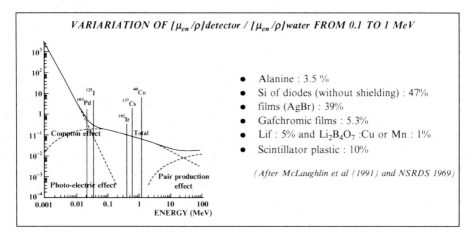

Figure 1. Solid state detectors irradiated with photon sources.

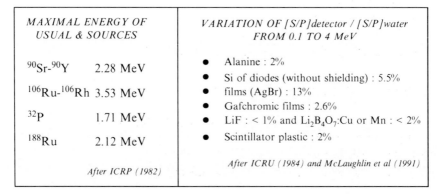

Figure 2. Solid state detectors irradiated with beta sources.

2. Diodes

2.1. PRINCIPLE

Most diodes are made of silicon that is a crystalline material where atomic electrons are arranged in bands of energy. During irradiation, electron–hole pairs are created. Conduction occurs by movement of electrons in the conduction band and by motion of holes in the valence band. The gap between the valence and the conduction band is large enough to prevent rapid thermal repopulation of electrons (or holes). The number of charge carriers is controlled by doping the material with impurities. Depending on the element used as impurity, the crystal is given either an excess or a deficit of free electrons that can carry electrical current. An excess of electrons carries a negatively charged current, giving a n-type semiconductor, whereas, an electron deficit carries a positively charged current giving a p-type semiconductor.[5]

Intracavitary available diodes dedicated to brachytherapy are generally mounted within flexible probes. They are made of a small quantity of silicon doped with phosphorus (n-type diodes) or boron (p-type diodes). In order to minimize leakage, they are operated without external bias voltage, their signals being measured either in open-circuit mode (voltage) or in short-circuit mode (current). This last mode is most often encountered in commercial electrometers and presents the advantage of producing a linear relationship between the charge generated in the diode and the dose. For detailed dosimetric properties and influence factors the reader can refer to "Handbook of Radiotherapy Physics".[15]

2.2. ADVANTAGES

- No external bias
- Immediate response
- Good intrinsic reproducibility ($\sigma < \pm 1\%$)
- Linear variation of response with dose up to 10 Gy

2.3. DISADVANTAGES

- Bad tissue-equivalence in the brachytherapy energy range[27]
- Temperature dependence
- Angular dependence
- Dose-rate and accumulated dose dependence

2.4. EXAMPLE OF APPLICATION

In vivo rectal dose measurements to avoid misadministrations during intra-cavitary HDR brachytherapy for cervix carcinoma.[1]

3. MOSFETs dosimeters

3.1. PRINCIPLE

Like diodes, MOSFET dosimeters belong to the semiconductor detectors. Two terminals of the MOSFET called the source and the drain are situated on top of a p-doped silicon region. A third terminal is the gate. Underneath the gate is an insulating silicon di-oxide layer and underneath this layer is the silicon substrate. The region of the substrate immediately below the oxide layer is known as the channel region. When a bias voltage is applied between source and drain, a sufficiently negative voltage – the "threshold voltage" – must be applied to the gate in order to allow a predetermined current to flow. During the irradiation of the MOSFET, a number of the holes produced are captured into traps close to the $Si-SiO_2$ interface. During the irradiation the gate is kept at a positive bias in order to enhance this phenomenon and thus to increase the sensitivity. The trapped holes cause a negative voltage shift in the threshold voltage, which is proportional to the number of trapped holes. This voltage shift can then be used for dosimetry, by measuring the threshold voltage before and after the irradiation.[28,31]

3.2. ADVANTAGES

- "Immediate" response, however not really "on-line", because after the end of the irradiation the gate voltage must be determined.
- No cable.
- Small size (of the order of 1 mm^3).
- Good reproducibility: 2–3% (1 σ).
- Response independent on dose rate.
- Negligible angular dependence ($\pm$ 2% for 360°).

3.3. DISADVANTAGES

- Response dependent on temperature[*].

---

[*] Both drawbacks can be eliminated by associating two detectors with a different grid voltage and by displaying the difference in their signals.

- Response decrease as a function of accumulated dose,[1] resulting in a limited lifetime of the detector (of the order of 150–200 Gy in the usual conditions).
- Bad tissue-equivalence in the brachytherapy energy range.[32]
- Slight loss of charge after irradiation: readings to be always made with the same time delay after the end of the irradiation.

3.4. EXAMPLES OF APPLICATION

Used as in vivo dosimeters to check the dwell positions of HDR ^{192}Ir source[30] and for in vivo dose monitoring during permanent ^{125}I interstitial implants.[10]

4. EPR-alanine

4.1. PRINCIPLE

The most common EPR dosimeter material is the amino acid L-α alanine. The interaction of ionizing radiation with this solid detector results in the formation of free radicals (molecules having one or more unpaired electrons). The induced transitions in the electron spin states are used to estimate the relative quantity of free radicals which depends on the beam quality and on the dose. It is evaluated by measuring the central peak height of the electron paramagnetic signal with a specific EPR spectrometer.[24] The readout procedure is non-destructive to the EPR signal.

4.2. ADVANTAGES

- Versatile presentation: small cylinders (4–5 mm in diameter and length varying from 1 to 30 mm) or films (0.25 mm in thickness) which can be cut in different shapes
- Accurate response ($\sigma = \pm 1\%$ for doses from 10 to 10^5 Gy)
- Stable response with time (fading 0.6% per year)
- EPR signal proportional to dose over a wide dose range (1–5,000 Gy)
- No accumulated dose dependence
- No dose-rate dependence up to 10^2 Gy/s and 10^8 Gy/s for continuous and pulsed radiations, respectively[20]
- Good tissue equivalence
- No directional effect

4.3. DISADVANTAGES

- Low sensitivity.
- Not well adapted for in vivo dosimetry (sensibility threshold 0.1–0.3 Gy).

- EPR spectrometer expensive and of difficult use but the dosimeters can be sent to a specialized laboratory for readout.

4.4. EXAMPLES OF APPLICATION

EPR/alanine dosimetry is currently used for HDR source calibration and mailed dose calibration services at several dosimetry laboratories worldwide.

5. Plastic scintillators

5.1. PRINCIPLE

The scintillator detectors consist of a base material (usually polystyrene, polyvinyl-toluene or acrylic-naphtalene) and one or more organic dyes embedded in it. When photons or charged particles pass through it, a few per cent of the energy loss is ultimately converted to scintillation photons. The light intensity is then either amplified and collected by photomultiplier tubes,[6] or focused on the photocathode of an image intensifier by a light guide and captured by a charge coupled digital camera.[26] In both of the cases, the plastic detector yields a count rate which can be correlated to the absorbed dose received by it.

5.2. ADVANTAGES

- Good tissue equivalence[9,12]
- Good reproducibility and stability
- Negligible dependence on temperature
- Linear response versus dose from some cGy to some Gy
- Independence on dose-rate
- High spatial resolution

5.3. DISADVANTAGES

- Precautions to be taken because of Cerenkov corrections in the lightguide.

5.4. EXAMPLE OF APPLICATION

Laboratory readers with multichannel PMTs offer possibilities to establish dose distribution around radioactive sources or within photon or electron beams.[4,12] One catheter based systems dedicated to the calibration of beta vascular sources and dose distribution around them are commercially available.

6. Radiochromic films

6.1. PRINCIPLE

The present possibilities offered by radiochromic films[31] are essentially due to improvement in the homogeneity and sensitivity of commercial emulsions. They consist of one or several sensitive layers made of microcrystals of radiation-sensitive monomers uniformly dispersed in a gelatin binder. Films are colorless and transparent before being irradiated. When they are exposed to ionizing radiation, polymerisation of the microcrystals occurs and the polymers alter the crystal colour to various shades of colour depending on the type of film (e.g. blue for Gafchromic films type MD 55, HS and EBT, and orange for type XR-T or RTQA).[†] The mechanism for radiation induced coloration has been demonstrated to be essentially due to a first order solid state polymerisation of the ordered substituted diacetylene monomers, which produce a main carbon chain planar conjugation.[21]

The colour change depends on the absorbed dose and its measurement is optimised through the use of instruments specially designed to measure the principal absorption peak in the spectrum of the used Gafchromic film.[8,11,13] Modern office flatbed scanners associated to commercial software can also give good results at less expensive cost.[2,3,29,33]

6.2. ADVANTAGES

- Self-developing (no physical, chemical or thermal processing)
- Very thin and easily cut to the desired shape and dimensions
- Tissue-equivalent over the energy range of brachytherapy[7]
- Good reproducibility from 2% to 5% (1 σ) with well defined procedure
- Non-destructive readout (possibility of dose accumulation over the different treatment sessions provided correction factors be introduced to take into account the variation of optical density with time)
- High spatial resolution
- Response independent on dose-rate from 0.08 to 80 Gy/min (Mc Laughlin *et al.* 1996)

[†] Details on composition and dosimetric properties of Gafchromic films can be found on: http:\\www.*ispcorp.com/products/dosimetry/index.htlm*.

6.3. DISADVANTAGES

- Reduced sensitivity (for instance response range for Gafchromic EBT is 0.1–40 Gy).
- No-linearity of dose response curve (but corrections easy with modern readout systems).
- Several hours necessary for the colour change to be sufficiently stabilized for evaluation. In practice it is recommended to read the films at constant delay after irradiation or to wait for a sufficient delay before readouts to get reproducible measurements.

6.4. EXAMPLES OF APPLICATION

Quality assurance for brachytherapy sources[8,14,23] and in vivo measurements for HDR ^{192}Ir applications.[25]

7.   Radiothermoluminescent dosimeters

7.1. PRINCIPLE

Thermoluminescence dosimetry[15,17] is based upon the ability of imperfect crystals to absorb and store the energy of ionizing radiation, which upon heating is re-emitted in the form of electromagnetic radiation, mainly in the visible wavelength. The light emitted, called thermoluminescence (TL) is then detected by a photomultipler (P.M.) and correlated to the absorbed dose received by the TL material. The emission thermoluminescence "spectrum" (or glow curve) consists of several TL peaks, corresponding to the different energy traps in the crystal. These include unstable peaks that must be eliminated by adequate pre-heating before readout and more stable peaks called dosimetric peaks that are the only ones that are useful for dosimetry. After readout, the TL material either returns to its original state or requires a special heating treatment called annealing in order to restore it.

TL detectors are either generated naturally or by doping phosphors with a very small percentage of activators.[18] They are available in the form of powder, solid dosimeters made of polycrystalline extrusions (rods, sintered pellets, or extremity chips). The TL material generally used in brachytherapy are different types of LiF, $Li_2B_4O_7$:Mn or $Li_2B_4O_7$:Cu which have the closest characteristics to those of tissues.[15]

Readers consist of the following components:

- **Heating system** housed in a **readout chamber**: Depending on the reader, it can be made of a metallic support (planchet) heated by an electric current or of isothermal fingers, or of hot nitrogen gas, or of an infrared sensor using an intense light pulse, or of a laser beam. It must offer the possibility to heat the TL dosimeter at two different temperatures: the preheating temperature used to clear off unstable peaks and the readout temperature used to collect the information from dosimetric peaks. The readout chamber must be continuously flushed with nitrogen gas in order to reduce oxygen linked spurious phenomena.[17]
- **Light detection system**: The luminous flux emitted by the TL dosimeter is collected and transmitted by a light guide into a photomultiplier (PM) tube with a bialkali photocathode. Optical filters are placed in front of the PM window in order to improve the response.
- **Signal integrator**: The signal proportional to the light emission is either amplified and feeds to an integrator (d.c. operation regimen) or converted into pulses and feeds to a scaler (pulse counting regimen). In most readers, glow curves are also displayed during dose measurements.
- **PC and associated software**: PC both controls the reading process and allows data management (dose calculations provided individual calibration factors are entered, statistical analysis, display and printout of experimental results...).

7.2. ADVANTAGES

- High sensitivity under a very small volume.
- Reproducibility of $\pm2\%$ or less can be obtained in routine use.
- Independence with dose rate up to $2.10^9\,\mathrm{Gy\ s^{-1}}$ (LiF, $Li_2B_4O_7Mn$ and CaF_2).
- Low fading.
- Absence of directional effect.
- Absence of variation with temperature in usual conditions.
- Low variation of the response as a function of energy of photon or electron emitters, provided the TL material be correctly chosen.

7.3. DISADVANTAGES

- Postponed response
- Destructive reading

7.4. EXAMPLES OF APPLICATION

TLD measurement of anisotropy factor used in most of brachytherapy computer programs[22] and in vivo measurements of the dose such as measurement of the dose delivered to the axillary zone during brachytherapy of breast tumors using ^{192}Ir.[16]

References

1. Alecu R. and Alecu M. (1999) In vivo rectal dose measurements with diodes to avoid mis-administrations during intracavitaryhigh dose rate brachytherapy for carcinoma of the cervix, Med. Phys. **26**, 768–770.
2. Alva H., Mercado-Uribe H., Rodriguez-Villafuerte M. and Brandan M.E. (2002) The use of a reflective scanner to study radiochromic film response. Phys. Med. Biol. **47**, 2925–2933.
3. Aydarous A.S., Darley P.J. and Charles M.W. (2001) A wide dynamic range, high spatial resolution scanning system for radiochromic dye films. Phys. Med. Biol. **46**, 1379–1389.
4. Bambinek M., Flühs D., Heintz M., Kolanoski H., Wegener D. and Quast U. (1999) Fluorescence 125I applicator. Med. Phys. **26**, 2476–2481.
5. Barthe J. (2001) Electronic dosimeters based on solid state detectors. Nucl. Instrum. Meth. Phys. Res., Sect. B, **184**, 158–189.
6. Beddar A.S., Mackie T.R. and Attix F.H. (1992) Water-equivalent plastic scintillation detectors for high energy beam dosimetry- 1. Physical characteristics and theoretical considerations. Phys. Med. Biol. **37**, 1883–1900.
7. Butson M.J., Cheung T. and Yu P. (2006) Weak energy dependence of EBT Gafchromic film dose response in the 50 kVp-10 MVp X-ray range. Appl. Radiat. Isotopes. **64**, 60–62.
8. Chui-Tsao S.-T., Duckworth T.L., Patel N.S., Pisch J. and Harrisson L.B. (2004) Verification of Ir-192 near source dosimetry using Gafchromic film. Med. Phys. **31**, 201–207.
9. Clift M.A., Sutton R.A. and Webb D.V. (2000) Water equivalence of plastic organic scintillators in megavoltage radiotherapy bremsstrahlung beams. Phys. Med. Biol. **45**, 1885–1895.
10. Cygler J., Saoudi A., Perry G., Morash C. and Choan E. (2006) Feasibility study of using MOSFET detectors for in vivo dosimetry during permanent low-dose rate prostate implants. Radiother. Oncol. **80**, 296–301.
11. Devic S., Seuntjens J., Heygil G., Podgorsak E.B., Soares C.G., Kirov A.S., Ali I., Williamson J. and Elizondo A. (2004) Dosimetric properties of improved Gafchromic films from seven digitizers. Med. Phys. **31**, 2392–2401.
12. Flühs D., Heintz M., Indekämpen F., Wieczorek C., Kolaoski H. and Quast U. (1996) Direct reading measurement of absorbed dose with plastic scintillators –The general concept and applications to ophtalmic plaque dosimetry, Med. Phys. **23**, 427–434.
13. Fusi F., Mercatelli L., Marconi G., Cuttone G. and Romano G. (2004) Optical characterization of radiochromic film by total reflectance and transmittance measurements. Med. Phys. **31**, 2147–2154.
14. Iftimia I., Devlin P.M., Chin L.M., Baron J.M. and Cormack R.A. (2003) GAF film dosimetry of a tandem of beta-emitting intravascular brachytherapy source train, Med. Phys. **30**, 1004–1012.

15. Marinello G. (2007) Radiothermoluminescent dosimeters and diodes. In "Hanbook of radio-therapy Physics" Eds Mayles P., Rosenwald J.C. and Nahum A. CRC Press-Taylor & Francis Group (London), chapter 16, 303–320.

16. Marinello G., Raynal M., Brulé A.M. and Pierquin B. (1975) Utilisation du fluorure de lithium en dosimétrie clinique. Application à la mesure de la dose délivrée à la région axillaire par l'iridium 192 dans l'endocuriethérapie des cancers du sein. J. Radiol. Elect. **11**, 791–796.

17. McKeever S.W.S. (1985) Thermoluminescence of solids. University Press (New York).

18. McKeever S.W.S., Moscovitch M. and Townsend P.D. (1995) Thermoluminescence dosi-metry materials: properties and uses. Nuclear Technology Publishing (Asford).

19. McLaughlin W.L., Boyd A.W., Chadwick K.H., Mc Donald J.C. and Miller A. (1989) Dosimetry for radiation processing. Taylor & Francis (London).

20. McLaughlin W.L., Chen Y-D. and Soares C.G. (1991) Sensitometry of the response of a new radiochromic film dosimeter to gamma and electron beams. Nucl. Instrum. Nad. Meth. Phys. Res. **A 302**, 165–176.

21. McLaughlin W.L., Al-Shiekley M., Lewis D.F., Kovacs A. and Wojnarovits L. (1996) A radiochromic solid state polymerization. In "Radiation effects on polymers" Eds Clough R.L., O'Donnel J.H. and Shalaby. American Chemical Society (Washington, DC), 152–166.

22. Meigooni A.S., Sanders M.F., Ibott G.S. and Szeglin S.R. (1997) Dosimetric characteristics of en imporved radiochromic film. Med. Phys. **23**, 1883–1886.

23. Mourtada F.A., Soares C.G., Seltzer S.M. and Lott S.H. (2000) Dosimetry characterization of ^{32}P catheter-based vascular brachytherapy source wire. Med. Phys. **27**, 1770–1776.

24. Olsen K.J., Hansen J.W. and Wille M. (1990) Response of the Alanine radiation dosimeterto high energy photon and electron beams. Phys. Med. Biol. **35**, 43–52.

25. Pai S., Reinstein L.E., Gluckman G., Xu Z. and Weiss T. (1998) The use of improved radio-chromic films for in vivo quality assurance of high dose rate brachytherapy. Med. Phys. **27**, 1217–1221.

26. Perera H., Williamson J.F., Monthofer S.P., Binns W.R., Klarmann J., Fuller G.L. and Wong J.W. (1992) Rapid-two dimensional dose-measurement in brachytherapy using plastic scintillator sheet: linearity, signal-to-noise ratio, and energy response characteristics. Int. J. Radiat. Oncol. Biol. Phys. **23**, 1059–1069.

27. Piermattei A., Azario L., Monaco G., Sorini A. and Arcovito G. (1995) p-type silicon detector for brachytherapy dosimetry. Med. Phys. **22**, 835–839.

28. Soubra M., Cygler J. and Mackay (1994) Evaluation of a dual bias dual metal oxide-silicon semiconductor field effect transistor detector as a radiation dosimeter. Med. Phys. **21**, 567–572.

29. Stevens M.A., Turner J.R., Hugtenburg R.P. and Butler P.H. (1996) High resolution dosimetry using radiochromic film and a document scanner. Phys. Med. Biol. **41**, 2357–2365.

30. Vassy D., Hallil A., Stubbs J., Webster M., Turmel J. and Salazar B. (2005) Verifying correct location of HDR Dwell position in the Mammosite catheter using an integral linear MOSFET dosimeter. Med. Phys. **32**, 1962.

31. Van Dam J. and Marinello G. (2006) Methods for in vivo dosimetry in external radiotherapy ESTRO Booklet No 1 (second edition). Available on : http:\\www. estro.org.

32. Wang B., Xu X.G. and Kim C.-H. (2005) Monte-Carlo study of MOSFET dosimeter char-acteristics: dose dependence on photon energy, direction and dosimeter composition. Radiat. Protect. Dosim., **113**, No 1, 40–46.

33. Yamauchi M., Tominaga T. and Nakamura O. (2004) Gafchromic film dosimetry with a flatbed color scanner for Leksell gamma knife therapy. Med. Phys. **31**, 1243–1248.

MONTE CARLO APPLICATION IN BRACHYTHERAPY DOSIMETRY

JOSE PEREZ-CALATAYUD[1],
DOMINGO GRANERO CABAÑERO[1],
FACUNDO BALLESTER PALLARÉS[2]
[1]*Physics Section. Radiotherapy Department,
La Fe University Hospital, Valencia, Spain
jperezc@uv.es*
[2]*Department of Atomic, Molecular and Nuclear Physics,
University of Valencia, Spain*

Abstract This paper devoted to Monte-Carlo Applications in BT Dosimetry is organized in four parts:

1. Motivation of the use of Monte Carlo in brachytherapy
2. The use of Monte Carlo to obtain dose distribution around brachytherapy sources
3. Experimental dosimetry versus Monte Carlo
4. Others Monte Carlo applications in brachytherapy

Keywords: PTRAN; ETRAN; ITS; MCNP; PENELOPE; GEANT4; EGS4; TG-43 U1 formalism; LTLE; TPS

1. Motivation of the use of Monte Carlo in brachytherapy

Experimental measurements are the base of brachytherapy (BT) dosimetry because all other calculated quantities by other methods must be validated against measurements. Nevertheless, the use of experimental methods in BT is very complicated because there is a high dose gradient around BT sources due to the inverse square law dominance. For example, the dose at 2 mm from the source is 25 times the dose at 1 cm. So, if we want to limit the dose uncertainty due to positioning errors to, for example, 2%, then the position uncertainty of the detectors on the phantom should be less than 50 and 20 μm at 5 and 2 mm from the source respectively.[44] Moreover, experimental methods in BT require to take into account too many detector aspects: linearity of response with dose, spatial resolution, energy dependence, influence of positioning errors in the experimental

Y. Lemoigne and A. Caner (eds.), *Radiotherapy and Brachytherapy*,
© Springer Science + Business Media B.V. 2009

set-up, sensitivity, volume averaging artefacts, etc. No one single dosimeter meets completely all the requirements. Today, the most popular detectors in BT are termoluminescent dosimeters (TLD) and radiochromic films.

By other hand, modern Monte Carlo (MC) codes use accurate cross section tables[17] and incorporate sophisticated tools to model BT sources and applicators geometry. So, MC has become a powerful tool for medical physics applications in general[20,36] and in BT dosimetry in particular.[38] MC-based dosimetry has been acknowledged as a valuable tool in brachytherapy, constituting one of the dosimetry prerequisites for routine clinical use of new low and high energy photon sources.[5]

Medical physicists have applied different MC codes to BT: PTRAN, ETRAN, ITS, MCNP, PENELOPE, GEANT4, EGS4, etc. The description that follows is largely based on the GEANT3[7] and GEANT4[1] codes although most of the codes implement similar features.

2. The use of Monte Carlo to obtain dose distribution around brachytherapy sources

In this section we are going briefly to describe the currently most important application of MC in BT: to obtain dose rate distribution in water around BT sources.

Treatment Planning Systems (TPS) calculations in BT are based upon the principle of source superposition assuming that dose rate distributions around BT sources present cylindrical symmetry, all medium is water and attenuation and inter-source effects are not taken into account. TPS obtain dose rate distribution around a source by means of interpolations over tabulated datasets (DRT). These DRT are used in TPS as input data and also to verify its calculations. Tables are expressed in rectangular coordinates (traditionally named *along and away*) or following the TG-43 U1 formalism 34.

The main steps to obtain these DRT using MC methods are the following:

1. Source geometry and material composition description
2. Phantoms characteristics and scoring voxels
3. Photon transport in water
4. Photon transport in air
5. Dose rate distribution in water
6. Along-away tables and TG-43 U1 data

2.1. SOURCE DESCRIPTION AND MATERIAL COMPOSITION DESCRIPTION

The basic information to model the source is obtained from the technical manufacturer's drawing and specifications. In some papers the authors perform auto- and radiography to verify these technical specifications of the source. The MC geometry tools are based on Boolean operations as intersections, unions, etc. with basic shapes that allows modeling accurately any source. Figure 1 shows an example of source geometry implementation for a Cs-137 source using different volume Boolean operations. Even the eyelet and its wedged end of this CSM3 source could be described. Once defined the geometry, composition materials are given by its fraction by weight and density.

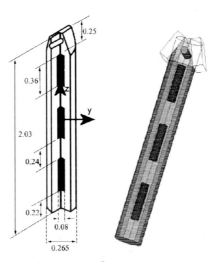

Figure 1. Source modelling with GEANT3[7] for the CSM3 Cs-137 source from BEBIG.

2.2. PHANTOMS CHARACTERISTICS AND SCORING VOXELS

The next step is to consider the water phantom to be used in the simulations. Usually, it is a water sphere with a diameter that varies from 30 cm for I-125 seeds to 100 cm for Co-60 sources although cylindrical phantoms were also used.[4,13,30]

The real trend is to use spherical phantoms that assure to obtain DRT under full scatter conditions up to the maximum distance of interest.[24,27]

Usually, the MC method uses voxels in which it is possible to score a lot of information as: number photons that cross the cell, its origin (primary or scatter), energy spectrum, kerma and absorbed dose, etc. Figure 2 shows an example of cell definition to obtain 2-D rectangular dose rate tables: the cells have 0.5 by 0.5 mm size and taking profit of the cylindrical symmetry around the longitudinal source axis the cell is defined as a ring around the longitudinal source axis, in order to get better statistics.

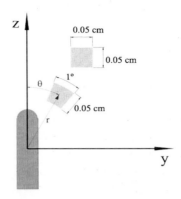

Figure 2. Cell definition, with cylindrical and spherical coordinates used in several studies.[4,13,30]

Voxel size is an important issue in BT studies (in both MC and measurements) because averaging dose in the voxel is assumed to be equal to the dose at the centre of the voxel.[3] This is not true in BT because dose is not a linear function with distance. This assumption produces a systematic error. To minimize it voxel sizes must be reasonable small.

2.3. PHOTON TRANSPORT IN WATER

With total cross sections MC code calculates the photon interaction mechanism in each collision and the distance between collisions. With the differential cross sections scattering angle and energy is randomized.

In principle, it is appropriate to include electron transport in the MC simulation. However this reduces the efficiency. To estimate absorbed dose, different variance reduction techniques have been developed to reduce the computing time required to achieve the desired level of statistical precision. Some of the variance reduction techniques are based on: (1) scoring kerma instead absorbed dose; (2) counting the number of photons that cross a cell or a surface; (3) measuring of the track in the cell; or (4) assigning a probability at each photon to the contribution in the calculation cell.[39]

So, some authors approximate absorbed dose by kerma but it must keep in mind that it is true when electronic equilibrium exists and it can not be applied where it doesn't exists, as for example, near the source (about 1–2 mm for Ir) and near interface media. A popular kerma estimator is the *linear track-length* (LTLE). Scoring the path length of the photon in the cell, L, kerma could be estimated by[39]:

$$K_V = \frac{L(\mu_{ab}/\rho)E_\gamma}{V} \tag{1}$$

where V is the cell volume.

2.4. AIR KERMA STRENGTH CALCULATION

Because the DRT are normalized to the air kerma strength, S_K, two separately MC simulations for each BT source are required:

Source located at the centre of a liquid-water phantom (usually a sphere), in order to obtain the dose rate distribution per contained activity.

Source located in vacuum or large air volume (a sphere several meters in diameter), to obtain the air kerma strength per contained activity.

Using both simulations dose rate distribution in water per unit of air kerma strength are obtained. Authors use different methods to estimate air kerma: (1) simulating photon transport in dry air and correcting air kerma by attenuation and scatter, or (2) simulated photon transport in vacuum storing photon fluence and spectra and estimating air kerma analytically. It has been shown that both methods give the same results.[18]

As an example of method (1) a volume $6 \times 6 \times 6$ m^3 of dry air is considered and the air kerma is scored along transversal source axis (from 5 to 150 cm) with cell size of 5×5 mm^2. The reference kerma is defined removing the scatter and attenuation in air. It is done with these two expressions due to Williamsom[41-43] for high and low energy sources

$$\begin{cases} k_{Air}(y)y^2 = S_K(1+\alpha y+\beta y^2)e^{-\mu_{Air}y} & \text{(low energy)} \\ \\ k_{Air}(y)y^2 = S_K(1+by) & \text{(high energy)} \end{cases} \tag{2}$$

For high energy, at each point the geometrical dependence is removed and these values are fitted to the linear equation. The slope b describes the deviation from inverse-square law due to the build-up of scattered photons in air and the intercept is an estimate of the reference air kerma. Similarly occurs for low energy expression. As example, in Fig. 3a, b the fit is show for a high energy source (Ir-192) and low energy one (I-125).

With this normalization of the DRT to the air kerma strength, the dose rate data in water are converted to dose rate per unit reference air kerma rate. Absorbed dose rates in water are obtained in the 2D rectangular table classically named along-away form and TG-43 functions, to be used on BT TPS.

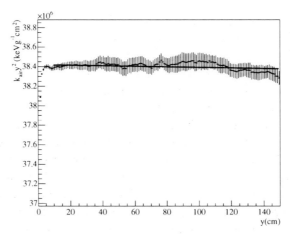

Figure 3a. Example of fit to obtain SK for high energy.

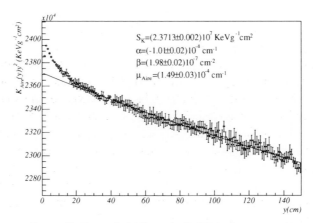

Figure 3b. Example of fit to obtain SK for low energy.

3. Experimental dosimetry *versus* Monte Carlo

The main disadvantages of experimental dosimetry *versus* MC calculations are the uncertainty in the positioning of the detectors near of the source (distances less than 1 cm form the source) due to the high dose gradient that appear near of the sources, and the poor signal-noise ratio for great distances from the source that affects the accuracy. By the other hand, the weak points that can affect to MC accuracy are: the knowledge of the right geometric configuration of the source, the uncertainties in the cross sections and the modelling of the physical processes in the MC code.

The optimal choice is to take advantages of each one. Making measurements in a set of points where the experimental uncertainties are low to verify MC calculations. After that, the validated MC code provides in detail all necessary dosimetry data around the source. For example, in I-125 seeds measure TLD measurements from 1–5 cm from the source can be used to validate MC calculations and then get all data with MC.

Other important aspect of the process is how to validate or to make the comparison of MC with measurements. For non-low energy radionuclides as Cs-137 and Ir-192, it is accepted that MC results can be validated comparing it with experimental results for a source of the same radionuclide and similar design.[21, 34] But for low energy (I-125, Pd-103) it is recommended that the validation must be done for the same source model[34] because the dependence on the source geometry and materials is very significant. AAPM recommends that for iodine and palladium seeds at least two independent dosimetric characterisations including experimental and MC studies must be performed for each seed model.

A verification of the geometric data of the sources provided by the manufacturer is recommended.[34] This is done by different methods. The most extended one is contact radiography. More sophisticated methods use techniques with special cameras or a microscopy.[19,41,43] Sometimes the verification show differences with respect to the manufacturer technical specifications of the source or deviations from cylindrical symmetry. The problem is that with contact radiography it is difficult to obtain fine details of the source. In practice, the problem of an accurate knowledge of the source geometry and materials is more or less solved when the study includes some experimental measurements. Then, if MC agrees with the experimental measurements, a realistic geometry has been considered.

Another important problem arises from the uncertainties in the geometry of the sources where tolerances in geometry can affect the dosimetry.[26,33,43] It is usual for low energy seeds, being the more frequent problem the internal movement of the active core inside of the capsule. So results are provided for an averaged source assuming no movements of the internal components.

There are a lot of source models on BT (around 70) each one with a specific design which implies a specific dosimetry for each one. The fundamental data of a source used in BT is the dose rate table in water around it. In Perez-Calatayud et al.[30] there is an example of the general process to get DRT for a source with a MC code experimentally validated.

In 2004 the AAPM[34] published a report (TG43 U1) directed to the recommendation of a set of dosimetric parameters of low energy sources by the use of MC or/and experimental dosimetry for its use in a clinical environment. For the experimental studies the TG43 U1 report states that all the experimental detectors are good for relative dosimetry but for absolute dosimetry recommends the use of TLD chips with a size of about $1 \times 1 \times 1$ mm^3. The studies using TLD must include the systematic (or type B) uncertainties (spatial resolution of the detectors, response of the detector with the energy, correction factors for water to solid medium (mainly water solid)) and statistical or type B uncertainties (that can be evaluated by means of repeated measurements). With respect to MC, the report states that all MC results must be verified using experimental methods and must include the type B uncertainties (in this case the uncertainty in the source geometry, the possible internal mobility of the source components, the uncertainties arising for the manufacturing process of the sources, the uncertainties in the knowledge of the atomic cross sections and the decay yield of the source and the uncertainties arising form the modelling of the physical process in the MC codes) and also the type A or statistical uncertainties.

In the TG43 U1 report of the AAPM a set of recommendations are given for the use of MC for low energy sources:

- The phantom used in the MC simulation must be a 30 cm in diameter water sphere.
- The statistical uncertainty must be ≤2% (for r < 5 cm) and ≤1% in S_k with $k = 1$.
- Modern cross sections libraries must be used equivalent to the NIST X-COM as DLC-146 or EPDL97. The use of older cross sections libraries could produce errors of up to 5–10% in low energy sources.[6]
- The geometry design and material composition of the source as provided by the manufacturer must be verified. Also the movement of the internal components of the source and the variation of the geometry of a set of sources must be studied and taken into account in the MC study.
- The difference between accumulate dose in a finite cell size instead of in a point (dose is a point function) must be less than 1%.
- The F(r, θ) calculation requires high resolution in zones with a high gradient of dose, as near of the source.
- A simulation of the WAFAC (Wide Angle Free Air Chamber), that is the chamber used in NIST for the calibration of I-125 and Pd-103 sources, must be mandatory in some cases in order to compare the obtained S_k with the MC code with that established in the NIST.

In the AAPM TG43 U1 report of 2004[34] and in a recent supplement of this report[35] a set of consensus datasets are established for low energy sources for its use in the clinical dosimetry. This consensus datasets have been obtained combining appropriately experimental and MC data and studying its compatibility between both datasets. If the MC and experimental datasets are compatibles within the uncertainties, the data chosen as consensus dataset is that with a large radial range, with more spatial resolution and the more smoothness data, being selected more frequently the MC data.

For the dose rate constant, Λ, the consensus value is the average of the MC and experimental value. For the radial dose function, g(r), and the anisotropy function the consensus values are the MC ones if they are compatible with the experimental data within the experimental uncertainties.

The AAPM in collaboration with ESTRO created a working group (HEBD-WG)[21] directed to the sources with a mean energy greater than 50 keV (mainly Cs-137 and Ir-192). The recommendations of this task group are similar to that of the AAPM TG43 U1 report with two exceptions: one in the way of validating the MC and another in the way of merge MC and experimental studies to obtain consensus datasets. The conclusions of this working group are that for high energy sources is enough to validate MC with a source of a similar design to the particular source in study and that the MC study is the selected for its use as a consensus dataset. The study for taking consensus dataset for Cs-137 and Ir-192 sources is currently in progress by the HEBD-WG.

4.   Others Monte Carlo applications in brachytherapy

Currently the principal application of MC in brachytherapy is directed to the calculation of dose rate tables around brachytherapy sources for its use in the TPS. There are several other applications where the MC method could be useful in brachytherapy.

It can be used for the study and design of applicators. This is the case, by example, of the vaginal cylinders with variable shielding[22] in that the transmission of the shielding could be evaluated and also the dose reduction due to the reduction of the scatter due to the shielding. Also there are MC studies of the colpostats with bladder-rectum shielding.[23] For the case of superficial applicators, MC is a necessary tool for its design and its dosimetry, as for example the Leipzig applicators[31] or the Valencia applicators.[15]

The MC method has been used also to evaluate the approximations in the calculations done by the TPS. An example of this is the application of the superposition principle, that is, to neglect the shielding of a source in another caused by the presence of various sources in LDR brachytherapy, as for example, in the Curietron,[11] Selectron[28] and in vaginal applicators of dome type.[32]

Another significant example of the use of MC in brachytherapy is its application in ocular brachytherapy with plaques[12]: by means of MC is possible to optimize the corrections factors used by the TPS.

MC is also applied in aspects of radiological protection in BT, as for example the evaluation of dose levels in the door of a bunker for HDR,[29] or in the calculation of the transmission curves in lead and concrete for different radioisotopes.[14]

Other interesting application of MC is the production of datasets to be used as input in the new TPS algorithms that allow correcting by the presence of a tissue of particular and by the effect in the change of the scatter component of the dose. These algorithms are currently in investigation by various groups[2,10,16,25,37,40] and are based in a parameterisation of the primary and scatter component of the dose, due to this dose rate tables with the primary and scatter components are needed and this is easily obtained with the use of MC.

Finally, the most actual application of MC is in the TPS in which a straight-forward MC calculation is done directly using the CT data of the patient. This is especially useful in low energy, as for example in the prostate implants in that the effect caused due to the presence of a tissue is very significant, the presence of calcifications in the prostate can lead to a reduction of up to 37% in the D90.[9] Some MC direct applications based on the CT prostate data of a patient have been developed by several groups. Williamson's group[9] has done the calculation of an implant with 97 seeds with a calculation time of less than 1 min. Others groups have also developed fast MC calculation engines for its use in the prostate implants.[8, 45]

References

1. Agostinelli S., Allison J. and Amako K., et al. "Geant4 – A Simulation Toolkit" Nucl. Ins. Meth. A 506. 250–303 (2003).
2. Anagnostopoulos G., Baltas D., Karaiskos P., Pantelis E., Papagiannis P. and Sakelliou L. "An analytical dosimetry model as a step towards accounting for inhomogeneities and bounded geometries in ^{192}Ir brachytherapy treatment planning" Phys. Med. Biol. 48. 1625–1647 (2003).
3. Ballester F., Hernandez C., Pérez-Calatayud J. and Lliso F. "Monte Carlo calculation of dose rate distributions around Ir-192 wires" Med. Phys. 24. 1221–1228 (1997).
4. Ballester F., Granero D., Perez-Calatayud J., Casal E. and Puchades V. "Monte Carlo dosimetric study of Best Industries and Alpha Omega Ir-192 brachytherapy seeds" Med. Phys. 31. 3298–3305 (2004).
5. Baltas D., Sakelliou L. and Zamboglou N., Chapter 9: Monte Carlo-Based Source Dosimetry, in The Physics of Modern Brachytherapy for Oncology Oncology, Taylor & Francis 2007.
6. Bohm T. D., DeLuca P. M. and DeWerd L. A. "Brachytherapy dosimetry of ^{125}I and ^{103}Pd sources using an updated cross section library for the MCNP Monte Carlo transport code" Med. Phys. 30. 701–711 (2003).
7. Brun R., Bruyant F., Marie M., McPherson A. C. and Zanarini P. GEANT3, CERN DD/EE/84-1 (1987).
8. Carrier J. F., D'Amours M., Verhaegen F., Reniers B., Martin A. G., Vigneault E. and Beaulieu L. "Postimplant dosimetry using a monte carlo dose calculation engine: A new clinical standard" Int. J. Radiat. Oncol. Biol. Phys. 68. 1190–1198 (2007).
9. Chibani O. and Williamson J. F. "MCPI: A sub-minute Monte Carlo dose calculation engine for prostate implants" Med. Phys. 32. 3688–3698 (2005).

10. Daskalov G. M., Baker R. S., Rogers D. W. O. and Williamson J. F. "Multigroup discrete ordinates modeling of 125I 6702 seed dose distributions using a broad energy-group cross section representation" Med. Phys. 29. 113–124 (2002).

11. Granero D., Puchades V., Pérez-Calatayud J., Ballester F. and Casal E. "Cálculo por Monte Carlo de la distribución de la tasa de dosis alrededor de la fuente de Cs-137 CSM1" Revista de Física Médica 5. 32–37 (2004).

12. Granero D., Pérez-Calatayud J., Ballester F., Casal E. and de Frutos J. M. "Dosimetric study of the 15 mm ROPES eye plaque" Med. Phys. 31. 3330–3336 (2004).

13. Granero D., Pérez-Calatayud J., Casal E., Ballester F. and Venselaar J. "A dosimetric study on the Ir-192 HDR Flexisource" Med. Phys. 33. 4578–4582 (2006).

14. Granero D., Pérez-Calatayud J., Ballester F., Bos A. and Venselaar J. "Broad-beam transmission data of new brachytherapy sources, Tm-170 and Yb-169" Radiat. Protect. Dosim. 118. 11–15 (2006).

15. Granero D., Pérez-Calatayud J., Gimeno. J, Ballester F., Casal E., Crispin V. and van der Laarse R. "Design and evaluation of a HDR skin applicator with flattening filter" Med. Phys. 35(2) (2008). Pendiente de número de paginas definitivo.

16. Hedtjarn H., Carlsson G. A. and Williamson J. F. "Accelerated Monte Carlo based dose calculations for brachytherapy planning using correlated sampling" Phys. Med. Biol. 47. 351–376 (2002).

17. Hubbell J. H. "Review and history of photon cross section calculations" Phys. Med. Biol. 51. R245–R262 (2006).

18. Karaiskos P., Angelopoulos A., Pantelis E., Papagiannis P., Sakelliou L., Kouwenhoven E. and Baltas D. "Monte Carlo dosimetry of a ^{192}Ir pulsed dose rate brachytherapy source" Med. Phys. 30. 9–16 (2003).

19. Kirov A. S., Meigooni A. S., Zhu Y., Valicenti R. K. and Williamson J. F. "Quantitative verification of ^{192}Ir PDR and HDR source structure by pin-hole autoradiography" Med. Phys. 22. 1753–1757 (1995).

20. Kling A., Barao F., Nakagawa M., Távora L. and Vaz P. (Eds.) Advanced Monte Carlo for Radiation Physics, Particle Transport Simulation and Applications, Proceedings of the Monte Carlo 2000 Conference, Lisbon, 23–26 October 2000, Springer, 2001.

21. Li Z., Das R. K., DeWerd L., Ibbot G. S., Meigooni A. S., Perez-Calatayud J., Rivard M. J., Sloboda R. S. and Williamson J. F. "Dosimetric prerequisites for routine clinical use of photon emitting brachytherapy sources with average energy higher than 50 KeV" Med. Phys. 34. 37–40 (2007).

22. Lymperopoulou G., Pantelis E., Papagiannis P., Rozaki-Mavrouli H., Sakelliou L., Baltas D. and Karaiskos P. "A Monte Carlo dosimetry study of vaginal ^{192}Ir brachytherapy applications with a shielded cylindrical applicator set" Med. Phys. 31. 3080–3086 (2004).

23. Markman J., Williamson J. F., Dempsey J. F. and Low D. A. "On the validity of the superposition principle in dose calculations for intracavitary implants with shielding vaginal colpostats" Med. Phys. 28. 147–155 (2001).

24. Melhus C. S. and Rivard M. J. "Approaches to calculating AAPM TG-43 brachytherapy dosimetry parameters for ^{137}Cs, ^{125}I, ^{192}Ir, ^{103}Pd, and ^{169}Yb sources" Med. Phys. 33. 1729–1737 (2006).

25. Pantelis E., Papagiannis P., Anagnostopoulos G., Baltas D., Karaiskos P., Sandilos P. and Sakelliou L. "Evaluation of a TG-43 compliant analytical dosimetry model in clinical ^{192}Ir HDR brachytherapy treatment planning and assessment of the significance of source position and catheter reconstruction uncertainties" Phys. Med. Biol. 49. 55–67 (2004).

26. Perera H., Williamson J., Li Z., Mishra V. and Meigooni A. "Dosimetric characteristics, air-kerma strength calibration and verification of the Monte Carlo simulation for a new Yterbium-169 brachytherapy source" Int. J. Radiat. Oncol. Biol. Phys., 28. 953–971 (1994).

27. Pérez-Calatayud J., Granero D. and Ballester F. "Phantom size in brachytherapy source dosimetric studies" Med. Phys. 31. 2075–2081 (2004).
28. Pérez-Calatayud J., Granero D., Ballester F., Puchades V. and Casal E. "Monte Carlo dosimetric characterization of the Cs-137 selectron/LDR source: Evaluation of applicator attenuation and superposition approximation effects" Med. Phys. 31. 493–499 (2004).
29. Pérez-Calatayud J., Granero D., Ballester F., Casal E., Crispin V., Puchades V., León A. and Verdú G. "Monte Carlo evaluation of kerma in an HDR brachytherapy bunker" Phys. Med. Biol. 49. N389–N396 (2004).
30. Pérez-Calatayud J., Granero D., Casal E., Ballester F. and Puchades V. "Monte Carlo and experimental derivation of TG43 dosimetric parameters for CSM-type Cs-137 sources" Med. Phys. 32. 28–36 (2005).
31. Pérez-Calatayud J., Granero D., Ballester F., Puchades V., Casal E., Soriano A. and Crispín V. "A dosimetric study of Leipzig applicators" Int. J. Radiat. Oncol. Biol. Phys. 62. 579–584 (2005).
32. Pérez-Calatayud J., Granero D., Ballester F. and Lliso F. "A Monte Carlo study of intersource effects in dome-type applicators loaded with LDR Cs-137 sources" Radiother. Oncol.. 77. 216–219 (2005).
33. Rivard M. J. "Monte Carlo calculations of AAPM Task Group Report No. 43 dosimetry parameters for the MED3631-A/M ^{125}I source" Med. Phys. 28. 629–637 (2001).
34. Rivard M. J., Coursey B. M., DeWerd L. A., Hanson W. F., Huq M. S., Ibbot G. S., Mitch M. G., Nath R. and Williamson J. F. "Update of the AAPM Task Group No 43 Report: A revised AAPM protocol for brachytherapy dose calculations" Med. Phys. 31. 633–674 (2004).
35. Rivard M. J., Butler W. M, DeWerd L. A., Huq M. S., Ibbott G. S., Meigooni A. S., Melhus C. S., Mitch M. G., Nath R. and Williamson J. F. "Supplement to the 2004 update of the AAPM Task Group No. 43 Report" Med. Phys. 34. 2187–2205 (2007).
36. Rogers D. W. O., "Fifty years of Monte Carlo simulations for medical physics" Phys. Med. Biol. 51. R287–R301 (2006).
37. Russell K. and Aneshjo A. "Dose calculation in brachytherapy for a 192Ir source using primary and scatter dose separation technique". Phys. Med. Biol. 41. 1007–1024 (1996).
38. Thomadsen B. R., Rivard M. J. and Butler W. M. (Eds.) Brachytherapy Physics. Second Edition. Proceedings of the Joint AAPM and ABS Summer School. Medical Physics Publishing, 2005.
39. Williamson J. F. "Monte Carlo evaluation of kerma at a point for photon transport problems," Med. Phys. 14. 567–576 (1987).
40. Williamson J. F., Perera H., Li Z. and Lutz W. R. "Comparison of calculated and measured heterogeneity correction factors for ^{125}I, ^{137}Cs, and ^{192}Ir brachytherapy sources near localized heterogeneities" Med. Phys. 20. 209–222 (1993).
41. Williamson J. and Li Z. "Monte Carlo aided dosimetry of the microSelectron pulsed and high dose-rate ^{192}Ir sources." Med. Phys. 22. 809–819 (1995).
42. Williamson J. F. "Monte Carlo modelling of the transverse-axis dose of the Model 200 ^{103}Pd interstitial brachytherapy source" Med. Phys. 27. 643–654 (2000).
43. Williamson J. F. "Dosimetric characteristics of the DRAXIMAGE model LS I-125 interstitial brachytherapy source design: A Monte Carlo investigation" Med. Phys. 29. 509–521 (2002).
44. Williamson J. F. in "Brachytherapy Physics. Second Edition". Joint AAPM/ABS Summer School. Thomadsen B., Rivard M. J., and Butler W. (Eds.), July 22–25, 2005, Seattle University, Seattle, WA
45. Yegin G., Taylor R. and Rogers D. "BrachyDose: a new fast Monte Carlo Code for brachytherapy calculations" Med. Phys. 33. 2074 (2006).

APPENDIX

GÜNTHER H. HARTMANN

Accuracy Strategies in Radiotherapy (11–27)

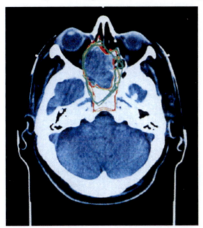 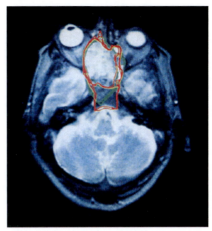

Figure 10. Four contours as outlined by four observers; contours in red outlined on CT image (right); contours in green outlined on axial MRI image (left).

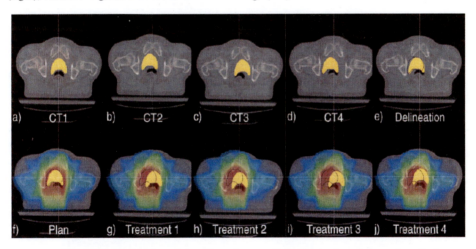

Figure 12. Simulated visualization of the effect of uncertainties on the dose delivery in radiotherapy. (a) Assume that the indicated region is the "true" clinical target volume (CTV), i.e., the region with tumor cells. The (white) cross wires indicate the room lasers. (b–d) Due to setup error on the CT scanner and organ motion, the four CT scans are not identical. (e) CT4 is used for planning, and the delineation (black contour) adds some extra error, because the "true" CTV is invisible. (f) The planning is based on the delineated CTV. (g–i) For each treatment fraction, the error made in the treatment preparation phase is reproduced, causing a systematic shift of the "true" CTV relative to the delivered dose distribution. In addition, random movements (treatment execution variations) occur due to tumor motion and setup error (Picture taken from [11]).

RAYMOND MIRALBELL

Clinical Applications Of 3-D Conformal Radiotherapy (87–94)

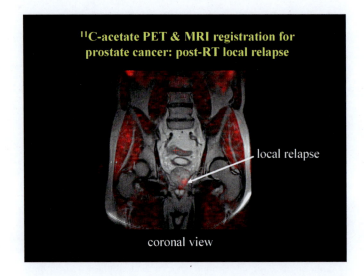

Figure 2. Example of prostate cancer.

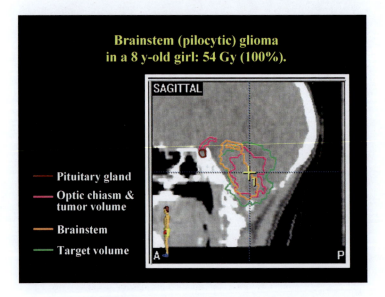

Figure 3a. Example of child brain-stem tumor.

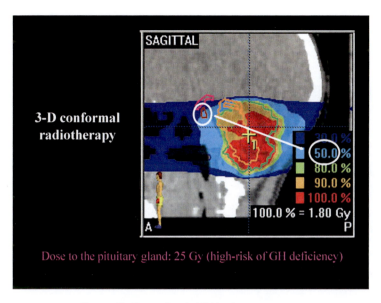

Figure 3b. Example of child brain-stem tumor.

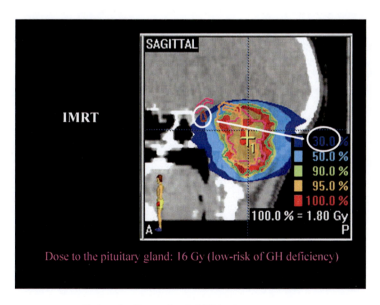

Figure 3c. Example of child brain-stem tumor.

SIMEON NILL

Treatment Planning: IMRT Optimization – Basic And Advanced Techniques (95–106)

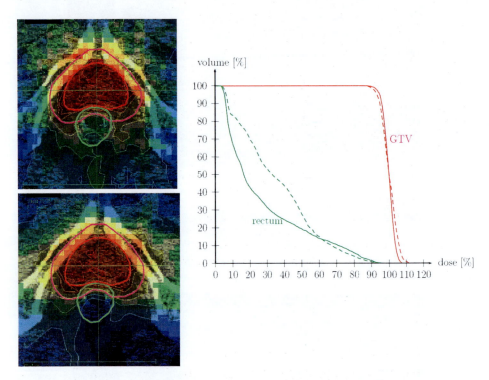

Figure 6. Comparison between a 6MV X field IMRT plan (top row) with an AMCBT plan (bottom row, 15). The solid lines in the DVH represent the AMCBT plan. Courtesy of Silke Ulrich.

PETER REMEIJER

Clinical Use Of 3-D Image Registration (127–133)

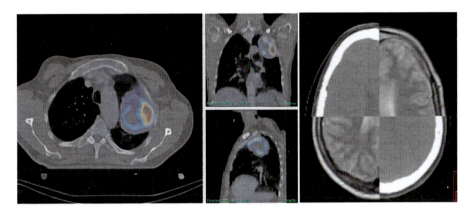

Figure 3. Examples of image fusion. The left image shows PET-CT fusion. The PET signal is shown as a color overlay on top of the CT scan. This works well because PET is a low resolution image (only used as an indicator to know if, e.g., a lymph node is positive) and geographical information is taken from the CT image. The right image shows 'sliding window' fusion of a CT and MR scan.

PETER REMEIJER

Geometrical Unceratinties in Radiotherapy (initial page–final page number)

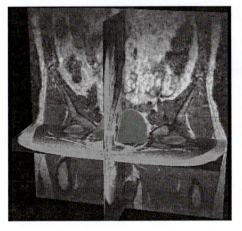 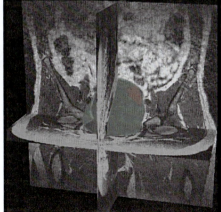

Figure 3. Organ motion. Depicted are two 3-D views of the bladder, taken one hour apart. The total movement of the cranial bladder wall was 7 cm, which shows that in some cases organ motion will be very difficult to determine a priori.

UWE OELFKE

Proton Therapy (173–181)

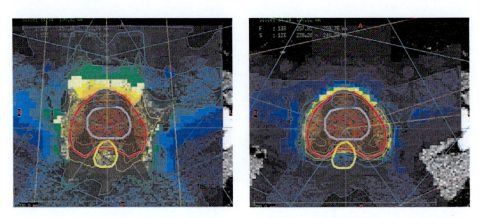

Figure 6. Treatment plan comparison for a prostate patient: left (IMRT photons); right (IMPT protons). On a transversal CT slice the dose distributions are shown as an overlay on the segmented anatomy of the patient. The shown volumes of interest refer to: prostate (light blue), extended target volume (red), bladder (dark blue) and rectum (yellow).

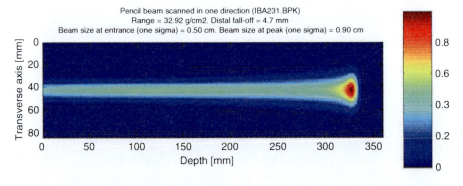

Figure 4. Relative dose distribution of a proton pencil beam in water.